Name _____

What's the Place?

Write the value of the underlined digit.

A. 2,351 __300__   4,825 _____   3,427 _____

B. 9,876 _____   5,820 _____   1,053 _____

C. 8,412 _____   1,357 _____   4,207 _____

D. 3,067 _____   8,165 _____   9,356 _____

E. 1,592 _____   9,706 _____   4,592 _____

Write the number.

F. 5 thousands 6 hundreds 1 ten 4 ones __5,614__

G. 6 thousands 8 hundreds 7 tens 1 one _____

H. two thousand eight hundred fifty-seven _____

I. 9 thousands 6 hundreds 0 tens 8 ones _____

J. one thousand five hundred seventy-four _____

K. 1 thousand 9 hundreds 2 tens 6 ones _____

L. two thousand three hundred sixty-one _____

M. 8 thousands 1 hundred 4 tens 2 ones _____

N. seven thousand six hundred thirteen _____

O. eight thousand twenty-seven _____

P. two thousand one _____

Q. 3 thousands 7 hundreds 5 tens 9 ones _____

R. four thousand seven hundred fifty _____

FS-32070 Third Grade Math Review

A Builder's Dream

1	2	3	4	5	6	7	8	9

Use the numbers above to write a four-digit number on each line.

A. the greatest number 9,876 _____

B. the greatest number with 7 in the thousands place _____

C. the least number _____

D. the greatest odd number _____

E. the least odd number _____

F. the greatest number with all even digits _____

G. the least number with all odd digits _____

Write the number or numbers that have the given value.

3,584   8,426   9,075   7,531   2,001

H. greater than 4,000 7,531;   8,426;   9,075

I. less than 2,500 _____

J. between 7,000 and 8,999 _____

K. less than 5,000 _____

L. between 1,000 and 2,999 _____

M. greater than 9,000 _____

FS-32070 Third Grade Math Review

Puzzle Place

Complete the puzzle by writing the correct number for each clue.

Across

A. six hundred fourteen thousand
   two hundred ninety-seven

E. three hundred twenty-two

F. five hundred twenty-nine

G. four thousand eighteen

H. twenty

J. ten

K. twenty-seven

L. forty-seven

M. nineteen thousand
   six hundred thirty-four

P. seven thousand
   eight hundred twenty

R. seven thousand
   four hundred thirty-five

S. sixty-two thousand
   five hundred thirty-one

Down

A. six hundred twenty-one

B. one hundred twenty-eight

C. nine hundred fifty-two

D. seven hundred twenty

E. thirty thousand ninety-four

G. four thousand one hundred
   seventeen

I. eight thousand seven
   hundred one

K. twenty-four

L. four hundred twenty-three

N. sixty-three

O. thirty-five

P. seventy-two

Q. eighty-five

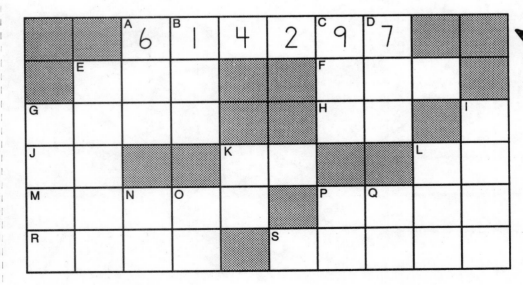

Name_____

Soaring With Numbers

Write the value of the underlined digit.

A. 4<u>6</u>8,930 _60,000_

B. 1<u>6</u>,497 _____

C. 5<u>4</u>,082 _____

D. <u>2</u>31,856 _____

E. 84<u>2</u>,031 _____

F. <u>6</u>02,937 _____

G. <u>8</u>69,235 _____

H. 2<u>3</u>7,981 _____

I. 4<u>9</u>,625 _____

J. <u>9</u>86,417 _____

84,<u>2</u>61 _____

<u>3</u>75,829 _____

49<u>3</u>,650 _____

58<u>7</u>,465 _____

<u>9</u>8,714 _____

850,<u>6</u>74 _____

9<u>2</u>6,504 _____

679,8<u>1</u>5 _____

<u>4</u>98,024 _____

5<u>2</u>3,127 _____

Write the numbers from the kites where the digit 6 has the given value.

K. 6,000 _____

L. 600 _____

M. 60,000 _____

N. 600,000 _____

260,357

842,623

693,712

436,129

4

Round About

Round to the nearest ten.

A. 29 _____30_____   48 _____   82 _____

B. 17 _____   31 _____   63 _____

C. 92 _____   59 _____   23 _____

D. 88 _____   12 _____   94 _____

E. 26 _____   37 _____   72 _____

F. 56 _____   66 _____   74 _____

G. 45 _____   88 _____   39 _____

Round to the nearest hundred.

H. 225 _____200_____   493 _____

I. 151 _____   290 _____

J. 412 _____   675 _____

K. 568 _____   946 _____

L. 178 _____   347 _____

M. 204 _____   763 _____

N. 915 _____   825 _____

O. 706 _____   891 _____

P. 174 _____   339 _____

Q. 887 _____   734 _____

R. 438 _____   510 _____

FS-32070 Third Grade Math Review

Rounding Numbers

Circle the numbers that would round to the first number in each row.

A. **80**	(76)	(83)	74	91	(75)	85
B. **40**	53	47	37	44	35	39
C. **60**	63	58	54	64	59	65
D. **50**	55	45	49	53	48	57
E. **30**	32	26	41	29	34	35
F. **90**	91	96	86	94	85	79

Circle the numbers that would round to the last number in each row.

G. (257)	352	403	(295)	(335)	**300**
H. 519	645	619	581	467	**500**
I. 875	914	850	950	923	**900**
J. 743	865	768	802	848	**800**
K. 193	137	98	143	129	**100**

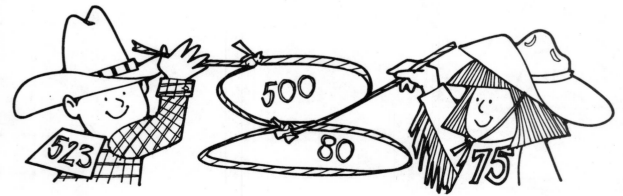

Catch the Facts

Add.

A.
5 + 7 12	7 + 9	8 + 4	6 + 9	9 + 4	8 + 9

B.
2 + 8	9 + 5	3 + 8	6 + 6	5 + 5	2 + 9

C.
9 + 8	6 + 4	4 + 7	9 + 7	3 + 9	6 + 8

D.
5 + 9	4 + 8	5 + 8	9 + 9	3 + 6	8 + 5

Write the missing number.

E. $7 + \boxed{8} = 15$ $4 + \square = 9$ $9 + \square = 12$ $6 + \square = 13$

F. $4 + \square = 10$ $6 + \square = 11$ $8 + \square = 14$ $4 + \square = 8$

G. $7 + \square = 13$ $3 + \square = 10$ $7 + \square = 14$ $8 + \square = 16$

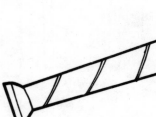  

     FS-32070 Third Grade Math Review

Name _____

Beat the Clock

Add.

A.
   5
+ 6

| |

   8
+ 7

   3
+ 4

   9
+ 6

   7
+ 9

   8
+ 5

B.
   6
+ 8

   9
+ 4

   5
+ 9

   8
+ 9

   4
+ 7

   3
+ 9

C.
   7
+ 3

   7
+ 8

   2
+ 7

   5
+ 7

   8
+ 8

   7
+ 5

D.
   6
+ 6

   4
+ 4

   5
+ 8

   7
+ 4

   2
+ 9

   8
+ 6

Write the missing number.

E. $9 + \boxed{5} = 14$ $6 + \Box = 15$ $6 + \Box = 13$

F. $3 + \Box = 10$ $4 + \Box = 10$ $9 + \Box = 16$

G. $6 + \Box = 10$ $3 + \Box = 11$ $7 + \Box = 13$

H. $8 + \Box = 12$ $9 + \Box = 17$ $6 + \Box = 11$

I. $9 + \Box = 18$ $7 + \Box = 14$ $4 + \Box = 13$

8

Over the Rainbow

Add.

A.
$$\begin{array}{r}1\\29\\+13\\\hline 42\end{array}$$
  $$\begin{array}{r}45\\+16\\\hline\end{array}$$
  $$\begin{array}{r}85\\+\ 9\\\hline\end{array}$$
  $$\begin{array}{r}53\\+18\\\hline\end{array}$$
  $$\begin{array}{r}42\\+29\\\hline\end{array}$$

B.
$$\begin{array}{r}45\\+39\\\hline\end{array}$$
  $$\begin{array}{r}57\\+18\\\hline\end{array}$$
  $$\begin{array}{r}69\\+12\\\hline\end{array}$$
  $$\begin{array}{r}48\\+38\\\hline\end{array}$$
  $$\begin{array}{r}26\\+26\\\hline\end{array}$$

C.
$$\begin{array}{r}37\\+18\\\hline\end{array}$$
  $$\begin{array}{r}29\\+56\\\hline\end{array}$$
  $$\begin{array}{r}43\\+37\\\hline\end{array}$$
  $$\begin{array}{r}71\\+19\\\hline\end{array}$$
  $$\begin{array}{r}47\\+49\\\hline\end{array}$$

D.
$$\begin{array}{r}18\\+18\\\hline\end{array}$$
  $$\begin{array}{r}52\\+18\\\hline\end{array}$$
  $$\begin{array}{r}66\\+\ 7\\\hline\end{array}$$
  $$\begin{array}{r}49\\+15\\\hline\end{array}$$
  $$\begin{array}{r}28\\+63\\\hline\end{array}$$

E.
$$\begin{array}{r}48\\+17\\\hline\end{array}$$
  $$\begin{array}{r}46\\+26\\\hline\end{array}$$
  $$\begin{array}{r}55\\+25\\\hline\end{array}$$
  $$\begin{array}{r}73\\+17\\\hline\end{array}$$
  $$\begin{array}{r}37\\+15\\\hline\end{array}$$

F.
$$\begin{array}{r}72\\+18\\\hline\end{array}$$
  $$\begin{array}{r}24\\+66\\\hline\end{array}$$
  $$\begin{array}{r}63\\+17\\\hline\end{array}$$

G.
$$\begin{array}{r}25\\+27\\\hline\end{array}$$
  $$\begin{array}{r}43\\+19\\\hline\end{array}$$
  $$\begin{array}{r}39\\+28\\\hline\end{array}$$

A Quilt of Problems

Add.

A.
$$\begin{array}{r} \overset{1}{38} \\ +16 \\ \hline 54 \end{array}$$
  $$\begin{array}{r} 47 \\ +13 \\ \hline \end{array}$$
  $$\begin{array}{r} 29 \\ +29 \\ \hline \end{array}$$
  $$\begin{array}{r} 35 \\ +25 \\ \hline \end{array}$$
  $$\begin{array}{r} 13 \\ +49 \\ \hline \end{array}$$

B.
$$\begin{array}{r} 19 \\ +13 \\ \hline \end{array}$$
  $$\begin{array}{r} 17 \\ +18 \\ \hline \end{array}$$
  $$\begin{array}{r} 19 \\ +26 \\ \hline \end{array}$$
  $$\begin{array}{r} 28 \\ +27 \\ \hline \end{array}$$
  $$\begin{array}{r} 37 \\ +35 \\ \hline \end{array}$$

C.
$$\begin{array}{r} 46 \\ +46 \\ \hline \end{array}$$
  $$\begin{array}{r} 53 \\ +29 \\ \hline \end{array}$$
  $$\begin{array}{r} 27 \\ +45 \\ \hline \end{array}$$
  $$\begin{array}{r} 59 \\ +\,3 \\ \hline \end{array}$$
  $$\begin{array}{r} 28 \\ +24 \\ \hline \end{array}$$

D.
$$\begin{array}{r} 27 \\ +23 \\ \hline \end{array}$$
  $$\begin{array}{r} 37 \\ +18 \\ \hline \end{array}$$
  $$\begin{array}{r} 48 \\ +12 \\ \hline \end{array}$$
  $$\begin{array}{r} 59 \\ +\,6 \\ \hline \end{array}$$
  $$\begin{array}{r} 39 \\ +31 \\ \hline \end{array}$$

E.
$$\begin{array}{r} 47 \\ +33 \\ \hline \end{array}$$
  $$\begin{array}{r} 39 \\ +39 \\ \hline \end{array}$$
  $$\begin{array}{r} 59 \\ +17 \\ \hline \end{array}$$
  $$\begin{array}{r} 49 \\ +25 \\ \hline \end{array}$$
  $$\begin{array}{r} 35 \\ +57 \\ \hline \end{array}$$

F.
$$\begin{array}{r} 68 \\ +22 \\ \hline \end{array}$$
  $$\begin{array}{r} 18 \\ +15 \\ \hline \end{array}$$
  $$\begin{array}{r} 24 \\ +29 \\ \hline \end{array}$$
  $$\begin{array}{r} 59 \\ +14 \\ \hline \end{array}$$
  $$\begin{array}{r} 65 \\ +28 \\ \hline \end{array}$$

FS-32070 Third Grade Math Review

Nutty Problems

Add.

A.
```
  1
 228
+343
─────
 571
```
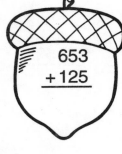
653
+125

318
+246

696
+202

224
+730

B.
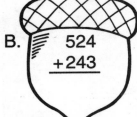
524
+243

425
+562

290
+199

614
+281

513
+777

C.
362
+537

279
+513

237
+421

697
+141

429
+239

D.
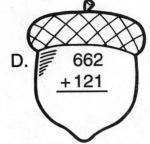
662
+121

544
+252

672
+772

123
+654

895
+442

E.
425
+125

308
+571

269
+521

810
+ 98

563
+172

F.
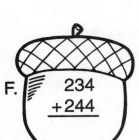
234
+244

705
+887

615
+193

68
+431

675
+119

Name _____

Going Bananas Over Addition

Add.

A.
233	763	617	135	90
+534	+152	+119	+531	+879
767				

B.
345	500	793	547	416
+190	+123	+142	+129	+ 83

C.
816	135	353	227	809
+147	+243	+122	+106	+ 90

D.
435	193	806	377	843
+534	+342	+ 82	+171	+ 25

E.
264	313	631	794	642
+462	+342	+136	+ 94	+319

F.
125	365	272
+101	+432	+416

G.
444	463	797
+190	+365	+202

FS-32070 Third Grade Math Review

Sum Path

Add.

A.   128
   + 113
   241

B.   431
   + 129

C.   170
   + 66

D.   367
   + 143

E.   183
   + 52

F.   494
   + 265

G.   327
   + 392

H.   368
   + 464

I.   379
   + 214

J.   842
   + 63

K.   766
   + 135

L.   543
   + 288

M.   450
   + 150

N.   372
   + 349

O.   175
   + 386

P.   674
   + 175

Q.   346
   + 195

R.   389
   + 102

S.   478
   + 135

T.   283
   + 149

Start at the toy store. Draw a line to connect the sums from problems A to T.

Toy Store	241	251	719	832	593	905
400	560	600	759	769	835	901
225	236	510	235	850	910	831
532	623	501	651	950	700	600
432	613	491	541	849	561	721

Exit

13

Name_____

All Together Now

Add.

I	U	F	G	D
11 235 +196 431	336 +365	264 +573	536 +156	836 + 47
W 236 +580	**E** 152 +356	**O** 676 +251	**N** 237 +628	**X** 299 +350
J 484 +292	**U** 321 +496	**K** 538 +191	**R** 214 +576	**G** 372 +244
R 903 + 48	**B** 793 +198	**H** 487 +305	**L** 679 +246	**O** 587 +406

Fill in the correct letter over each answer.
What is everything in the world doing at the same time?

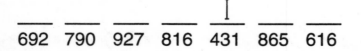

___ ___ ___ ___ I ___ ___   ___ ___ ___ ___ ___
692 790 927 816 431 865 616   993 925 883 508 951

14

Great Bear Sums

Add. Circle the greatest sum in each row.

A.   4,567   4,835   3,973   8,243
   + 1,360   + 4,124   + 1,024   +   961
   5,927

B.   5,832   3,467   3,009   6,253
   + 1,745   + 2,981   + 1,996   + 1,626

C.   6,175   4,862   5,032   3,125
   + 2,834   + 2,105   + 1,304   + 5,862

D.   5,152   6,371   2,484   1,396
   + 4,734   +   983   + 3,105   + 4,609

E.   2,166   2,156   8,653   9,137
   + 4,730   +   239   + 1,047   +   842

F.   2,786   8,197
   + 2,017   +   102

G.   6,854   8,266
   + 1,023   +   424

15   FS-32070 Third Grade Math Review

Name _____

Prickly Problems

Add.

A.
```
   1
  4,357        2,352        5,075        7,256
 +3,932       +6,435       +  123       +1,982
 ──────
  8,289
```

B.
```
  4,353        5,705        7,216        3,852
 +2,405       +1,836       +1,642       +3,975
```

C.
```
  5,036        6,852        2,834          972
 +2,823       +1,607       +5,165       +7,846
```

D.
```
  4,358        5,215        2,852        1,207
 +3,483       +2,732       +6,056       +3,402
```

E.
```
  5,697        7,285        6,343        5,216
 +1,903       +  614       +1,756       +1,893
```

F.
```
  4,367          842
 +2,332       +7,198
```

G.
```
  6,097        5,318
 +  903       +2,460
```

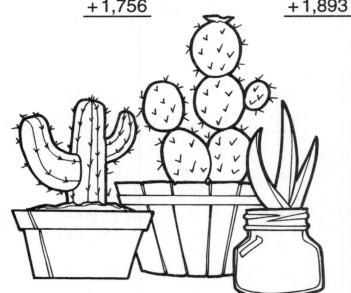

FS-32070 Third Grade Math Review

Triple Hitter

Add.

A.
```
  |
  45        28        70        81        53
  37        30        62        29        15
+ 14      + 19      + 57      + 43      + 22
  96
```

B.
```
  75        75        54        75        46
  63        28        95        19        97
+ 50      + 19      + 77      + 68      + 13
```

C.
```
  12        54        45        62        77
  89        17        32        78        86
+ 35      + 47      + 23      + 97      + 18
```

D.
```
  528       293       846       496       360
  306       687       797       185       875
+ 217     + 319     + 904     + 206     + 496
```

E.
```
  876       865       746
  257        79       829
+  18     + 491     + 688
```

F.
```
  826       496       221
  835        87       107
+ 206     + 142     + 322
```

Name _____

Shooting Stars

Add.

A.
```
   |
   75        29        88        58        23
   28        38        39        49        97
 +20       +27       +47       +26       +29
 ───
 123
```

B.
```
   24        69                  32        19
   66        12                  87        86
 +31       +14                 +43       +43
```

C.
```
             27        36        97        27
             50        45        35        18
           +38       +89       +22       +67
```

D.
```
  235       327       836                 297
  196       450       245                 635
 +267      +118      +589                +342
```

E.
```
  687                 830       490       865
  618                 103       374       231
 +167                +428      +221      +814
```

F.
```
  625       504       680       124
  103       190       208       179
 +134      +847      +112      +813
```

18

FS-32070 Third Grade Math Review

Totally Cool

Add.

A.
```
  |||
    347
 +2,693
  3,040
```
562
+975

3,652
+ 79

4,385
+1,823

B.
2,682
+1,937

972
+ 83

6,096
+ 997

3,862
+ 98

C.
4,975
+ 127

367
+2,896

4,065
+3,985

250
+475

D.
375
+987

2,856
+ 85

9,349
+ 467

1,126
+7,956

E.
2,223
+   85

537
+187

8,516
+1,095

2,987
+ 563

F.
8,675
+ 296

5,093
+1,923

6,781
+ 969

378
+699

FS-32070 Third Grade Math Review

"Sum"mer Fun

Add.

A.
```
   ı ı ı
  4,865
 +1,987
  ─────
  6,852
```

```
   132
  +989
  ────
```

```
  4,096
 +  875
  ─────
```

```
   328
  +597
  ────
```

B.
```
  3,117
 +  983
  ─────
```

```
  5,870
 +  562
  ─────
```

```
  1,835
 +6,926
  ─────
```

```
  8,458
 +   97
  ─────
```

C.
```
  4,497
 +2,536
  ─────
```

```
  1,432
 +  976
  ─────
```

```
  8,158
 +   87
  ─────
```

```
   858
  +596
  ────
```

D.
```
   386
  +509
  ────
```

```
  2,983
 +5,087
  ─────
```

```
  5,681
 +  299
  ─────
```

```
   943
  +876
  ────
```

E.
```
  6,888
 +  219
  ─────
```

```
  4,098
 +  923
  ─────
```

```
  6,948
 +2,065
  ─────
```

```
  3,342
 +2,964
  ─────
```

F.
```
  5,106
 +2,983
  ─────
```

```
   486
  +926
  ────
```

G.
```
  1,836
 +  890
  ─────
```

```
  8,817
 +  827
  ─────
```

20

In the Ballpark

Round to the nearest ten. Estimate the sum.

A.
32	30	79		56	
+68	+ 70	+67	+ ____	+39	+ ____
	100				

B.
81		17		73	
+58	+ ____	+46	+ ____	+66	+ ____

C.
78		62		43	
+43	+ ____	+91	+ ____	+48	+ ____

Round to the nearest hundred. Estimate the sum.

D.
342	300	625		297	
+287	+ 300	+303	+ ____	+502	+ ____
	600				

E.
569		749		685	
+119	+ ____	+475	+ ____	+804	+ ____

F.
483		858	
+915	+ ____	+743	+ ____

Name _____

Rounding Up and Down

Round to the nearest ten. Estimate the sum.

A. 43   40   69
   +28   +30   +73   + _____
          ———
            70

B. 47   21   54
   +63   + _____ +78   + _____ +93   + _____

C. 73   61   14
   +85   + _____ +56   + _____ +43   + _____

Round to the nearest hundred. Estimate the sum.

D. 507   500   889
   +635   +600   +330   + _____
          ———
           1,100

E. 625   582   873
   +254   + _____ +159 + _____ +198 + _____

                    F. 813   723
                       +569 + _____ +485 + _____

Dancing Differences

Subtract.

A.
$$\begin{array}{r} 12 \\ -\ 8 \\ \hline 4 \end{array}$$
$$\begin{array}{r} 15 \\ -\ 3 \\ \hline \end{array}$$
$$\begin{array}{r} 11 \\ -\ 3 \\ \hline \end{array}$$
$$\begin{array}{r} 10 \\ -\ 7 \\ \hline \end{array}$$
$$\begin{array}{r} 8 \\ -\ 4 \\ \hline \end{array}$$
$$\begin{array}{r} 9 \\ -\ 6 \\ \hline \end{array}$$

B.
$$\begin{array}{r} 16 \\ -\ 9 \\ \hline \end{array}$$
$$\begin{array}{r} 13 \\ -\ 7 \\ \hline \end{array}$$
$$\begin{array}{r} 15 \\ -\ 6 \\ \hline \end{array}$$
$$\begin{array}{r} 17 \\ -\ 8 \\ \hline \end{array}$$
$$\begin{array}{r} 14 \\ -\ 7 \\ \hline \end{array}$$
$$\begin{array}{r} 13 \\ -\ 4 \\ \hline \end{array}$$

C.
$$\begin{array}{r} 16 \\ -\ 8 \\ \hline \end{array}$$
$$\begin{array}{r} 14 \\ -\ 5 \\ \hline \end{array}$$
$$\begin{array}{r} 13 \\ -\ 6 \\ \hline \end{array}$$
$$\begin{array}{r} 16 \\ -\ 7 \\ \hline \end{array}$$
$$\begin{array}{r} 11 \\ -\ 8 \\ \hline \end{array}$$
$$\begin{array}{r} 10 \\ -\ 5 \\ \hline \end{array}$$

D.
$$\begin{array}{r} 18 \\ -\ 9 \\ \hline \end{array}$$
$$\begin{array}{r} 15 \\ -\ 7 \\ \hline \end{array}$$
$$\begin{array}{r} 14 \\ -\ 6 \\ \hline \end{array}$$
$$\begin{array}{r} 17 \\ -\ 9 \\ \hline \end{array}$$
$$\begin{array}{r} 10 \\ -\ 4 \\ \hline \end{array}$$
$$\begin{array}{r} 9 \\ -\ 4 \\ \hline \end{array}$$

Write the missing number.

E. $15 - \boxed{8} = 7$ $\qquad$ $10 - \boxed{} = 2$ $\qquad$ $12 - \boxed{} = 6$

F. $11 - \boxed{} = 6$ $\qquad$ $13 - \boxed{} = 5$ $\qquad$ $14 - \boxed{} = 8$

G. $10 - \boxed{} = 7$ $\qquad$ $12 - \boxed{} = 7$ $\qquad$ $12 - \boxed{} = 9$

H. $14 - \boxed{} = 9$ $\qquad$ $11 - \boxed{} = 5$ $\qquad$ $12 - \boxed{} = 8$

I. $12 - \boxed{} = 5$ $\qquad$ $13 - \boxed{} = 8$ $\qquad$ $13 - \boxed{} = 9$

Name _____

Subtraction Action

Subtract.

A.
13
− 9
4

10
− 3

16
− 8

10
− 6

11
− 6

12
− 6

B.
15
− 6

11
− 9

16
− 9

10
− 4

12
− 3

14
− 5

C.
18
− 9

17
− 9

11
− 8

12
− 4

13
− 8

12
− 7

D.
17
− 8

16
− 7

10
− 5

11
− 4

12
− 5

13
− 4

Write the missing number.

E. 10 − [2] = 8   15 − [ ] = 8   11 − [ ] = 4   12 − [ ] = 3

F. 11 − [ ] = 6   13 − [ ] = 6   13 − [ ] = 8   11 − [ ] = 8

G. 12 − [ ] = 4   14 − [ ] = 7   14 − [ ] = 6   10 − [ ] = 2

FS-32070 Third Grade Math Review

Clowning With Subtraction

Subtract.

A.
$$\begin{array}{r} {\scriptstyle 3\,10} \\ \cancel{40} \\ -12 \\ \hline 28 \end{array}$$
$$\begin{array}{r} 63 \\ -29 \\ \hline \end{array}$$
$$\begin{array}{r} 80 \\ -27 \\ \hline \end{array}$$
$$\begin{array}{r} 66 \\ -37 \\ \hline \end{array}$$
$$\begin{array}{r} 52 \\ -14 \\ \hline \end{array}$$

B.
$$\begin{array}{r} 93 \\ -77 \\ \hline \end{array}$$
$$\begin{array}{r} 95 \\ -56 \\ \hline \end{array}$$
$$\begin{array}{r} 63 \\ -11 \\ \hline \end{array}$$
$$\begin{array}{r} 83 \\ -\ 9 \\ \hline \end{array}$$
$$\begin{array}{r} 59 \\ -30 \\ \hline \end{array}$$

C.
$$\begin{array}{r} 68 \\ -52 \\ \hline \end{array}$$
$$\begin{array}{r} 71 \\ -\ 5 \\ \hline \end{array}$$
$$\begin{array}{r} 43 \\ -25 \\ \hline \end{array}$$
$$\begin{array}{r} 43 \\ -18 \\ \hline \end{array}$$
$$\begin{array}{r} 55 \\ -36 \\ \hline \end{array}$$

D.
$$\begin{array}{r} 81 \\ -12 \\ \hline \end{array}$$
$$\begin{array}{r} 94 \\ -76 \\ \hline \end{array}$$
$$\begin{array}{r} 85 \\ -27 \\ \hline \end{array}$$
$$\begin{array}{r} 96 \\ -38 \\ \hline \end{array}$$
$$\begin{array}{r} 83 \\ -22 \\ \hline \end{array}$$

E.
$$\begin{array}{r} 80 \\ -67 \\ \hline \end{array}$$
$$\begin{array}{r} 46 \\ -43 \\ \hline \end{array}$$
$$\begin{array}{r} 75 \\ -27 \\ \hline \end{array}$$
$$\begin{array}{r} 78 \\ -65 \\ \hline \end{array}$$
$$\begin{array}{r} 82 \\ -76 \\ \hline \end{array}$$

F.
$$\begin{array}{r} 59 \\ -42 \\ \hline \end{array}$$
$$\begin{array}{r} 32 \\ -18 \\ \hline \end{array}$$
$$\begin{array}{r} 56 \\ -48 \\ \hline \end{array}$$

G.
$$\begin{array}{r} 79 \\ -34 \\ \hline \end{array}$$
$$\begin{array}{r} 91 \\ -65 \\ \hline \end{array}$$
$$\begin{array}{r} 65 \\ -32 \\ \hline \end{array}$$

25

FS-32070 Third Grade Math Review

Name _____

Cuddly Subtraction

Subtract.

A.
```
   4 11
    5̸1̸
  − 18
   33
```
```
   62
 − 13
```
```
   83
 −  4
```
```
   71
 − 47
```
```
   76
 − 55
```

B.
```
   97
 − 27
```
```
   50
 − 18
```
```
   76
 − 29
```
```
   88
 − 36
```
```
   66
 − 57
```

C.
```
   72
 − 45
```
```
   95
 − 22
```
```
   33
 − 18
```
```
   59
 − 32
```
```
   62
 − 53
```

D.
```
   88
 − 32
```
```
   79
 − 20
```
```
   60
 − 19
```
```
   67
 −  9
```
```
   88
 − 79
```

E.
```
   50
 − 38
```
```
   66
 − 54
```
```
   83
 − 36
```
```
   43
 − 22
```
```
   80
 − 44
```

F.
```
   91
 − 27
```
```
   85
 − 43
```
```
   75
 − 65
```
```
   42
 −  8
```
```
   76
 − 43
```

FS-32070 Third Grade Math Review

Name _____

Carnival Math

Subtract.

A.
$$\begin{array}{r} 6\,11 \\ 6\cancel{7}\cancel{1} \\ -134 \\ \hline 537 \end{array}$$

$$\begin{array}{r} 243 \\ -\ 82 \\ \hline \end{array}$$

$$\begin{array}{r} 926 \\ -432 \\ \hline \end{array}$$

B.
$$\begin{array}{r} 764 \\ -138 \\ \hline \end{array}$$

$$\begin{array}{r} 134 \\ -\ 74 \\ \hline \end{array}$$

$$\begin{array}{r} 825 \\ -195 \\ \hline \end{array}$$

C.
$$\begin{array}{r} 328 \\ -295 \\ \hline \end{array}$$

$$\begin{array}{r} 692 \\ -\ 38 \\ \hline \end{array}$$

$$\begin{array}{r} 892 \\ -153 \\ \hline \end{array}$$

$$\begin{array}{r} 725 \\ -\ 92 \\ \hline \end{array}$$

$$\begin{array}{r} 683 \\ -456 \\ \hline \end{array}$$

D.
$$\begin{array}{r} 425 \\ -180 \\ \hline \end{array}$$

$$\begin{array}{r} 627 \\ -392 \\ \hline \end{array}$$

$$\begin{array}{r} 821 \\ -340 \\ \hline \end{array}$$

$$\begin{array}{r} 866 \\ -293 \\ \hline \end{array}$$

$$\begin{array}{r} 892 \\ -375 \\ \hline \end{array}$$

E.
$$\begin{array}{r} 835 \\ -362 \\ \hline \end{array}$$

$$\begin{array}{r} 952 \\ -825 \\ \hline \end{array}$$

$$\begin{array}{r} 378 \\ -293 \\ \hline \end{array}$$

$$\begin{array}{r} 610 \\ -301 \\ \hline \end{array}$$

$$\begin{array}{r} 842 \\ -290 \\ \hline \end{array}$$

F.
$$\begin{array}{r} 738 \\ -193 \\ \hline \end{array}$$

$$\begin{array}{r} 625 \\ -491 \\ \hline \end{array}$$

$$\begin{array}{r} 816 \\ -385 \\ \hline \end{array}$$

$$\begin{array}{r} 293 \\ -147 \\ \hline \end{array}$$

$$\begin{array}{r} 514 \\ -283 \\ \hline \end{array}$$

G.
$$\begin{array}{r} 773 \\ -444 \\ \hline \end{array}$$

$$\begin{array}{r} 628 \\ -575 \\ \hline \end{array}$$

$$\begin{array}{r} 275 \\ -\ 67 \\ \hline \end{array}$$

$$\begin{array}{r} 970 \\ -865 \\ \hline \end{array}$$

$$\begin{array}{r} 766 \\ -392 \\ \hline \end{array}$$

FS-32070 Third Grade Math Review

Name _____

Submerge Into Subtraction

Subtract.

A.
$$\begin{array}{r} {\scriptstyle 6\ 13} \\ \cancel{732} \\ -141 \\ \hline 591 \end{array}$$

$$\begin{array}{r} 244 \\ -\ 73 \\ \hline \end{array}$$

$$\begin{array}{r} 817 \\ -382 \\ \hline \end{array}$$

$$\begin{array}{r} 753 \\ -272 \\ \hline \end{array}$$

$$\begin{array}{r} 671 \\ -333 \\ \hline \end{array}$$

B.
$$\begin{array}{r} 819 \\ -584 \\ \hline \end{array}$$

$$\begin{array}{r} 135 \\ -\ 84 \\ \hline \end{array}$$

$$\begin{array}{r} 781 \\ -127 \\ \hline \end{array}$$

$$\begin{array}{r} 980 \\ -252 \\ \hline \end{array}$$

$$\begin{array}{r} 734 \\ -571 \\ \hline \end{array}$$

C.
$$\begin{array}{r} 438 \\ -229 \\ \hline \end{array}$$

$$\begin{array}{r} 644 \\ -526 \\ \hline \end{array}$$

$$\begin{array}{r} 249 \\ -168 \\ \hline \end{array}$$

$$\begin{array}{r} 384 \\ -\ 92 \\ \hline \end{array}$$

$$\begin{array}{r} 848 \\ -663 \\ \hline \end{array}$$

D.
$$\begin{array}{r} 776 \\ -285 \\ \hline \end{array}$$

$$\begin{array}{r} 641 \\ -350 \\ \hline \end{array}$$

$$\begin{array}{r} 338 \\ -253 \\ \hline \end{array}$$

$$\begin{array}{r} 162 \\ -\ 48 \\ \hline \end{array}$$

$$\begin{array}{r} 871 \\ -258 \\ \hline \end{array}$$

E.
$$\begin{array}{r} 982 \\ -374 \\ \hline \end{array}$$

$$\begin{array}{r} 625 \\ -431 \\ \hline \end{array}$$

$$\begin{array}{r} 474 \\ -\ 81 \\ \hline \end{array}$$

$$\begin{array}{r} 880 \\ -452 \\ \hline \end{array}$$

$$\begin{array}{r} 467 \\ -284 \\ \hline \end{array}$$

F.
$$\begin{array}{r} 463 \\ -181 \\ \hline \end{array}$$

$$\begin{array}{r} 861 \\ -425 \\ \hline \end{array}$$

$$\begin{array}{r} 767 \\ -249 \\ \hline \end{array}$$

$$\begin{array}{r} 627 \\ -275 \\ \hline \end{array}$$

$$\begin{array}{r} 935 \\ -341 \\ \hline \end{array}$$

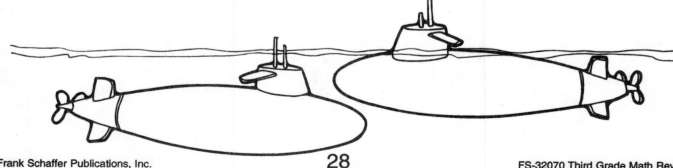

FS-32070 Third Grade Math Review

Pick of the Litter

Subtract.

A.
$$\begin{array}{r} \overset{15}{\overset{6\ \overset{\cancel{8}}{}\ 15}{\cancel{765}}} \\ -376 \\ \hline 389 \end{array}$$

814
−635

932
−894

843
−658

754
−559

B.
826
− 79

356
−277

777
−488

548
−269

695
−396

C.
752
−578

764
−288

635
−358

566
− 78

518
−329

D.
647
−259

747
− 59

456
− 78

841
−499

826
−358

E.
356
−197

934
−235

945
−297

718
−269

716
−497

F.
915
− 26

442
−384

567
−499

563
−274

354
−268

Name _____

Playful Problems

Subtract.

A.
$$
\begin{array}{r}
^{1}4\ ^{15}5\ ^{13}3 \\
\cancel{563} \\
-178 \\
\hline
385
\end{array}
\qquad
\begin{array}{r}862 \\ -495 \\ \hline\end{array}
\qquad
\begin{array}{r}713 \\ -198 \\ \hline\end{array}
\qquad
\begin{array}{r}326 \\ -129 \\ \hline\end{array}
\qquad
\begin{array}{r}553 \\ -\ 68 \\ \hline\end{array}
$$

B.
$$
\begin{array}{r}723 \\ -427 \\ \hline\end{array}
\qquad
\begin{array}{r}261 \\ -\ 87 \\ \hline\end{array}
\qquad
\begin{array}{r}546 \\ -379 \\ \hline\end{array}
\qquad
\begin{array}{r}743 \\ -295 \\ \hline\end{array}
\qquad
\begin{array}{r}983 \\ -597 \\ \hline\end{array}
$$

C.
$$
\begin{array}{r}764 \\ -466 \\ \hline\end{array}
\qquad
\begin{array}{r}793 \\ -195 \\ \hline\end{array}
\qquad
\begin{array}{r}541 \\ -166 \\ \hline\end{array}
\qquad
\begin{array}{r}552 \\ -278 \\ \hline\end{array}
\qquad
\begin{array}{r}824 \\ -295 \\ \hline\end{array}
$$

D.
$$
\begin{array}{r}353 \\ -268 \\ \hline\end{array}
\qquad
\begin{array}{r}777 \\ -378 \\ \hline\end{array}
\qquad
\begin{array}{r}964 \\ -175 \\ \hline\end{array}
\qquad
\begin{array}{r}915 \\ -496 \\ \hline\end{array}
\qquad
\begin{array}{r}821 \\ -722 \\ \hline\end{array}
$$

E.
$$
\begin{array}{r}121 \\ -\ 99 \\ \hline\end{array}
\qquad
\begin{array}{r}828 \\ -749 \\ \hline\end{array}
\qquad
\begin{array}{r}875 \\ -387 \\ \hline\end{array}
\qquad
\begin{array}{r}871 \\ -296 \\ \hline\end{array}
\qquad
\begin{array}{r}321 \\ -233 \\ \hline\end{array}
$$

F.
$$
\begin{array}{r}727 \\ -438 \\ \hline\end{array}
\qquad
\begin{array}{r}874 \\ -396 \\ \hline\end{array}
\qquad
\begin{array}{r}532 \\ -166 \\ \hline\end{array}
\qquad
\begin{array}{r}516 \\ -329 \\ \hline\end{array}
\qquad
\begin{array}{r}974 \\ -596 \\ \hline\end{array}
$$

FS-32070 Third Grade Math Review

Crystal Ball

Subtract. Fill in the correct letter over each answer.
What is it that you cannot see, but is always ahead of you?

			$\overset{\text{T}}{}$			
308	58	391	405	384	245	6	355	67

T	S	E	J	H
$\overset{7\ 10}{\cancel{8}\cancel{0}9}$ -564 $\overline{245}$	180 $-\ 42$	705 -314	430 -214	903 -845
B	**G**	**U**	**C**	**D**
510 $-\ 76$	805 -688	530 -146	603 -478	800 -496
U	**K**	**L**	**T**	**E**
304 -298	400 -291	900 -567	604 -296	700 -633
P	**M**	**F**	**N**	**R**
906 -609	800 -743	603 -198	408 -285	702 -347

FS-32070 Third Grade Math Review

Name _____

A Secret Word

Subtract. Color the boxes with differences of 500 or less.

A. $\begin{array}{r} 504 \\ -275 \\ \hline 229 \end{array}$	$\begin{array}{r} 905 \\ -386 \\ \hline \end{array}$	$\begin{array}{r} 700 \\ -267 \\ \hline \end{array}$	$\begin{array}{r} 806 \\ -259 \\ \hline \end{array}$	$\begin{array}{r} 703 \\ -255 \\ \hline \end{array}$
B. $\begin{array}{r} 508 \\ -369 \\ \hline \end{array}$	$\begin{array}{r} 908 \\ -\ 79 \\ \hline \end{array}$	$\begin{array}{r} 805 \\ -473 \\ \hline \end{array}$	$\begin{array}{r} 701 \\ -\ 48 \\ \hline \end{array}$	$\begin{array}{r} 800 \\ -246 \\ \hline \end{array}$
C. $\begin{array}{r} 307 \\ -268 \\ \hline \end{array}$	$\begin{array}{r} 440 \\ -328 \\ \hline \end{array}$	$\begin{array}{r} 805 \\ -485 \\ \hline \end{array}$	$\begin{array}{r} 904 \\ -265 \\ \hline \end{array}$	$\begin{array}{r} 506 \\ -175 \\ \hline \end{array}$
D. $\begin{array}{r} 706 \\ -380 \\ \hline \end{array}$	$\begin{array}{r} 650 \\ -\ 67 \\ \hline \end{array}$	$\begin{array}{r} 703 \\ -208 \\ \hline \end{array}$	$\begin{array}{r} 900 \\ -312 \\ \hline \end{array}$	$\begin{array}{r} 700 \\ -248 \\ \hline \end{array}$
E. $\begin{array}{r} 902 \\ -789 \\ \hline \end{array}$	$\begin{array}{r} 800 \\ -284 \\ \hline \end{array}$	$\begin{array}{r} 801 \\ -357 \\ \hline \end{array}$	$\begin{array}{r} 603 \\ -\ 27 \\ \hline \end{array}$	$\begin{array}{r} 804 \\ -399 \\ \hline \end{array}$
F. $\begin{array}{r} 706 \\ -407 \\ \hline \end{array}$	$\begin{array}{r} 607 \\ -\ 59 \\ \hline \end{array}$	$\begin{array}{r} 700 \\ -295 \\ \hline \end{array}$	$\begin{array}{r} 906 \\ -329 \\ \hline \end{array}$	$\begin{array}{r} 808 \\ -343 \\ \hline \end{array}$

What is the secret word? _____

Four-digit subtraction
with regrouping once

Soar Through the Air

Subtract.

A.
$$
\begin{array}{r}
6\;14\\
4,3\not7\not4\\
-2,256\\
\hline
2,118
\end{array}
$$

$$
\begin{array}{r}
6,636\\
-5,429\\
\hline
\end{array}
$$

$$
\begin{array}{r}
5,295\\
-2,883\\
\hline
\end{array}
$$

B.
$$
\begin{array}{r}
5,843\\
-3,561\\
\hline
\end{array}
$$

$$
\begin{array}{r}
8,254\\
-5,832\\
\hline
\end{array}
$$

$$
\begin{array}{r}
3,173\\
-1,158\\
\hline
\end{array}
$$

C.
$$
\begin{array}{r}
3,680\\
-1,342\\
\hline
\end{array}
$$

$$
\begin{array}{r}
5,663\\
-4,249\\
\hline
\end{array}
$$

$$
\begin{array}{r}
6,202\\
-4,051\\
\hline
\end{array}
$$

$$
\begin{array}{r}
3,768\\
-1,924\\
\hline
\end{array}
$$

D.
$$
\begin{array}{r}
6,170\\
-3,158\\
\hline
\end{array}
$$

$$
\begin{array}{r}
4,431\\
-2,630\\
\hline
\end{array}
$$

$$
\begin{array}{r}
2,846\\
-1,593\\
\hline
\end{array}
$$

$$
\begin{array}{r}
5,845\\
-1,629\\
\hline
\end{array}
$$

E.
$$
\begin{array}{r}
7,059\\
-5,343\\
\hline
\end{array}
$$

$$
\begin{array}{r}
8,868\\
-6,339\\
\hline
\end{array}
$$

$$
\begin{array}{r}
7,908\\
-3,150\\
\hline
\end{array}
$$

$$
\begin{array}{r}
5,291\\
-3,078\\
\hline
\end{array}
$$

F.
$$
\begin{array}{r}
8,367\\
-2,623\\
\hline
\end{array}
$$

$$
\begin{array}{r}
7,746\\
-5,385\\
\hline
\end{array}
$$

$$
\begin{array}{r}
8,692\\
-2,486\\
\hline
\end{array}
$$

$$
\begin{array}{r}
5,446\\
-1,329\\
\hline
\end{array}
$$

G.
$$
\begin{array}{r}
7,665\\
-2,754\\
\hline
\end{array}
$$

$$
\begin{array}{r}
9,456\\
-4,173\\
\hline
\end{array}
$$

$$
\begin{array}{r}
5,243\\
-1,823\\
\hline
\end{array}
$$

$$
\begin{array}{r}
3,543\\
-1,493\\
\hline
\end{array}
$$

FS-32070 Third Grade Math Review

Highflying Kites

Subtract.

A.
$$\begin{array}{r} 7\ 16 \\ 7,8\cancel{86} \\ -4,079 \\ \hline 3,807 \end{array}$$
$$\begin{array}{r} 4,374 \\ -2,861 \\ \hline \end{array}$$

B.
$$\begin{array}{r} 8,364 \\ -7,229 \\ \hline \end{array}$$
$$\begin{array}{r} 8,746 \\ -2,485 \\ \hline \end{array}$$

C.
$$\begin{array}{r} 6,234 \\ -1,520 \\ \hline \end{array}$$
$$\begin{array}{r} 6,670 \\ -6,341 \\ \hline \end{array}$$
$$\begin{array}{r} 9,538 \\ -9,275 \\ \hline \end{array}$$
$$\begin{array}{r} 9,327 \\ -1,257 \\ \hline \end{array}$$

D.
$$\begin{array}{r} 5,759 \\ -1,939 \\ \hline \end{array}$$
$$\begin{array}{r} 7,287 \\ -1,532 \\ \hline \end{array}$$
$$\begin{array}{r} 6,876 \\ -2,459 \\ \hline \end{array}$$
$$\begin{array}{r} 5,314 \\ -2,504 \\ \hline \end{array}$$

E.
$$\begin{array}{r} 6,805 \\ -5,472 \\ \hline \end{array}$$
$$\begin{array}{r} 7,620 \\ -3,280 \\ \hline \end{array}$$
$$\begin{array}{r} 5,832 \\ -5,160 \\ \hline \end{array}$$
$$\begin{array}{r} 6,473 \\ -4,392 \\ \hline \end{array}$$

F.
$$\begin{array}{r} 6,274 \\ -3,463 \\ \hline \end{array}$$
$$\begin{array}{r} 1,723 \\ -1,504 \\ \hline \end{array}$$
$$\begin{array}{r} 4,139 \\ -2,618 \\ \hline \end{array}$$
$$\begin{array}{r} 7,623 \\ -5,408 \\ \hline \end{array}$$

G.
$$\begin{array}{r} 6,924 \\ -3,452 \\ \hline \end{array}$$
$$\begin{array}{r} 5,296 \\ -4,167 \\ \hline \end{array}$$
$$\begin{array}{r} 9,423 \\ -5,418 \\ \hline \end{array}$$
$$\begin{array}{r} 7,258 \\ -4,735 \\ \hline \end{array}$$

Name _____

Mountain Climbing

Subtract.

A.
```
  11 14
 5 /X/13
  6,253
 -1,896
  4,357
```
1,537
− 875

3,731
− 879

6,208
−1,823

B.
4,055
−1,967

4,619
−2,682

6,955
−3,983

7,003
− 907

C.
3,960
−   98

9,325
−8,750

5,000
−3,975

6,913
−3,578

D.
8,050
−3,985

9,725
− 942

1,362
− 987

9,806
− 457

E.
6,208
−3,985

3,001
−1,503

8,012
−2,751

8,771
−   96

F.
4,723
−1,952

6,075
−2,738

G.
7,018
−1,923

9,007
−8,669

Name _____

Investigating

Subtract.

A.
$$\begin{array}{r} \overset{12}{\cancel{2}}\overset{\cancel{7}}{\,}\overset{10}{\,} \\ \cancel{3,305} \\ -1,391 \\ \hline 1,914 \end{array}$$
$$\begin{array}{r} 6,140 \\ -4,350 \\ \hline \end{array}$$
$$\begin{array}{r} 3,623 \\ -1,496 \\ \hline \end{array}$$
$$\begin{array}{r} 5,704 \\ -932 \\ \hline \end{array}$$

B.
$$\begin{array}{r} 6,200 \\ -172 \\ \hline \end{array}$$
$$\begin{array}{r} 1,405 \\ -67 \\ \hline \end{array}$$
$$\begin{array}{r} 8,240 \\ -5,639 \\ \hline \end{array}$$
$$\begin{array}{r} 8,047 \\ -6,239 \\ \hline \end{array}$$

C.
$$\begin{array}{r} 3,200 \\ -1,195 \\ \hline \end{array}$$
$$\begin{array}{r} 7,042 \\ -5,453 \\ \hline \end{array}$$
$$\begin{array}{r} 9,000 \\ -4,170 \\ \hline \end{array}$$
$$\begin{array}{r} 5,403 \\ -2,794 \\ \hline \end{array}$$

D.
$$\begin{array}{r} 6,088 \\ -295 \\ \hline \end{array}$$
$$\begin{array}{r} 4,896 \\ -1,497 \\ \hline \end{array}$$
$$\begin{array}{r} 5,000 \\ -1,818 \\ \hline \end{array}$$
$$\begin{array}{r} 8,075 \\ -79 \\ \hline \end{array}$$

E.
$$\begin{array}{r} 4,309 \\ -1,428 \\ \hline \end{array}$$
$$\begin{array}{r} 7,058 \\ -5,466 \\ \hline \end{array}$$
$$\begin{array}{r} 1,003 \\ -757 \\ \hline \end{array}$$
$$\begin{array}{r} 9,637 \\ -829 \\ \hline \end{array}$$

F.
$$\begin{array}{r} 9,009 \\ -567 \\ \hline \end{array}$$
$$\begin{array}{r} 1,600 \\ -1,189 \\ \hline \end{array}$$
$$\begin{array}{r} 8,083 \\ -2,989 \\ \hline \end{array}$$
$$\begin{array}{r} 6,022 \\ -5,926 \\ \hline \end{array}$$

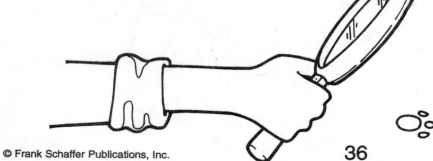

FS-32070 Third Grade Math Review

Best Estimate

Round to the nearest ten. Estimate the difference.

A. 38
   −12
   $\begin{array}{r} 40 \\ -10 \\ \hline 30 \end{array}$

   57
   −28   − _____

   81
   −49   − _____

B. 93
   −27   − _____

   41
   −19   − _____

   85
   −46   − _____

C. 53
   −31   − _____

   42
   −28   − _____

   92
   −68   − _____

Round to the nearest hundred. Estimate the difference.

D. 508
   −295
   $\begin{array}{r} 500 \\ -300 \\ \hline 200 \end{array}$

   789
   −315   − _____

E. 693
   −210   − _____

   414
   −298   − _____

F. 809
   −417   − _____

   785
   −302   − _____

FS-32070 Third Grade Math Review

Name _____ Estimating differences

A Pile of Gifts

Round to the nearest ten. Estimate the difference.

A.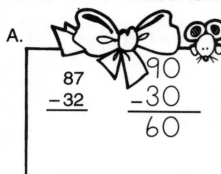

$$\begin{array}{r} 87 \\ -32 \\ \hline \end{array} \qquad \begin{array}{r} 90 \\ -30 \\ \hline 60 \end{array}$$

$$\begin{array}{r} 93 \\ -12 \\ \hline \end{array} \qquad \underline{}$$

$$\begin{array}{r} 46 \\ -28 \\ \hline \end{array} \qquad \underline{}$$

B.

$$\begin{array}{r} 58 \\ -29 \\ \hline \end{array} \qquad \underline{}$$

$$\begin{array}{r} 86 \\ -17 \\ \hline \end{array} \qquad \underline{}$$

$$\begin{array}{r} 51 \\ -42 \\ \hline \end{array} \qquad \underline{}$$

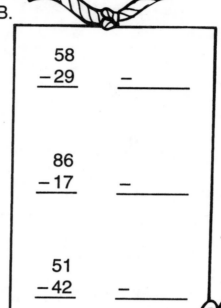

C.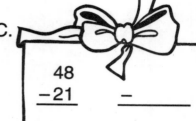

$$\begin{array}{r} 48 \\ -21 \\ \hline \end{array} \qquad \underline{}$$

$$\begin{array}{r} 75 \\ -47 \\ \hline \end{array} \qquad \underline{}$$

$$\begin{array}{r} 93 \\ -89 \\ \hline \end{array} \qquad \underline{}$$

Round to the nearest hundred. Estimate the difference.

D.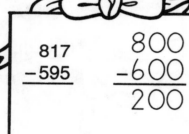

$$\begin{array}{r} 817 \\ -595 \\ \hline \end{array} \qquad \begin{array}{r} 800 \\ -600 \\ \hline 200 \end{array}$$

$$\begin{array}{r} 675 \\ -398 \\ \hline \end{array} \qquad \underline{}$$

$$\begin{array}{r} 780 \\ -594 \\ \hline \end{array} \qquad \underline{}$$

E.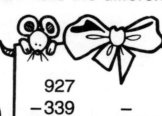

$$\begin{array}{r} 927 \\ -339 \\ \hline \end{array} \qquad \underline{}$$

$$\begin{array}{r} 514 \\ -209 \\ \hline \end{array} \qquad \underline{}$$

$$\begin{array}{r} 883 \\ -597 \\ \hline \end{array} \qquad \underline{}$$

F.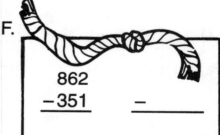

$$\begin{array}{r} 862 \\ -351 \\ \hline \end{array} \qquad \underline{}$$

$$\begin{array}{r} 732 \\ -109 \\ \hline \end{array} \qquad \underline{}$$

$$\begin{array}{r} 415 \\ -398 \\ \hline \end{array} \qquad \underline{}$$

FS-32070 Third Grade Math Review

Count by Groups

Write a number sentence for each picture.

A.

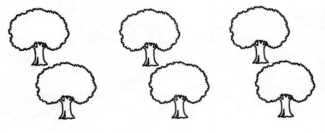

___3___ x ___2___ = ___6___   _____ x _____ = _____

B.

_____ x _____ = _____   _____ x _____ = _____

Multiply.

D.
$\begin{array}{r} 2 \\ \times\ 2 \\ \hline 4 \end{array}$
$\begin{array}{r} 7 \\ \times\ 5 \\ \hline \end{array}$
$\begin{array}{r} 1 \\ \times\ 5 \\ \hline \end{array}$
$\begin{array}{r} 3 \\ \times\ 2 \\ \hline \end{array}$
$\begin{array}{r} 5 \\ \times\ 2 \\ \hline \end{array}$
$\begin{array}{r} 4 \\ \times\ 5 \\ \hline \end{array}$

E.
$\begin{array}{r} 8 \\ \times\ 2 \\ \hline \end{array}$
$\begin{array}{r} 3 \\ \times\ 5 \\ \hline \end{array}$
$\begin{array}{r} 5 \\ \times\ 5 \\ \hline \end{array}$
$\begin{array}{r} 1 \\ \times\ 2 \\ \hline \end{array}$
$\begin{array}{r} 8 \\ \times\ 5 \\ \hline \end{array}$
$\begin{array}{r} 4 \\ \times\ 2 \\ \hline \end{array}$

F.
$\begin{array}{r} 9 \\ \times\ 5 \\ \hline \end{array}$
$\begin{array}{r} 6 \\ \times\ 5 \\ \hline \end{array}$
$\begin{array}{r} 6 \\ \times\ 2 \\ \hline \end{array}$
$\begin{array}{r} 9 \\ \times\ 2 \\ \hline \end{array}$
$\begin{array}{r} 2 \\ \times\ 5 \\ \hline \end{array}$
$\begin{array}{r} 7 \\ \times\ 2 \\ \hline \end{array}$

Name _____

Multiplication Snakes

Multiply.

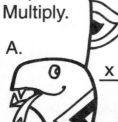

A.
3	5	8	4	3	9
x 2	x 2	x 2	x 5	x 5	x 2
6					

B.
4	6	1	5	2	8
x 2	x 2	x 5	x 5	x 2	x 5

C.
7	2	9	6	1	7
x 2	x 5	x 5	x 5	x 2	x 5

Draw a picture to show each number sentence.
Then write the product.

D.   7 x 2 = __14__   4 x 2 = _____

◯ ◯ ◯ ◯ ◯ ◯ ◯
◯ ◯ ◯ ◯ ◯ ◯ ◯

E.   3 x 2 = _____   4 x 5 = _____

40

Name _____

Multiplication March

Multiply.

A.
$\begin{array}{r} 5 \\ \times\ 3 \\ \hline 15 \end{array}$
$\begin{array}{r} 4 \\ \times\ 4 \\ \hline \end{array}$
$\begin{array}{r} 7 \\ \times\ 3 \\ \hline \end{array}$
$\begin{array}{r} 4 \\ \times\ 3 \\ \hline \end{array}$
$\begin{array}{r} 2 \\ \times\ 4 \\ \hline \end{array}$
$\begin{array}{r} 3 \\ \times\ 1 \\ \hline \end{array}$

B.
$\begin{array}{r} 6 \\ \times\ 3 \\ \hline \end{array}$
$\begin{array}{r} 1 \\ \times\ 3 \\ \hline \end{array}$
$\begin{array}{r} 3 \\ \times\ 4 \\ \hline \end{array}$
$\begin{array}{r} 8 \\ \times\ 4 \\ \hline \end{array}$
$\begin{array}{r} 9 \\ \times\ 3 \\ \hline \end{array}$
$\begin{array}{r} 7 \\ \times\ 4 \\ \hline \end{array}$

C.
$\begin{array}{r} 9 \\ \times\ 4 \\ \hline \end{array}$
$\begin{array}{r} 2 \\ \times\ 3 \\ \hline \end{array}$
$\begin{array}{r} 4 \\ \times\ 6 \\ \hline \end{array}$
$\begin{array}{r} 5 \\ \times\ 4 \\ \hline \end{array}$
$\begin{array}{r} 3 \\ \times\ 5 \\ \hline \end{array}$
$\begin{array}{r} 4 \\ \times\ 2 \\ \hline \end{array}$

D.
$\begin{array}{r} 8 \\ \times\ 3 \\ \hline \end{array}$
$\begin{array}{r} 1 \\ \times\ 4 \\ \hline \end{array}$
$\begin{array}{r} 3 \\ \times\ 3 \\ \hline \end{array}$
$\begin{array}{r} 6 \\ \times\ 4 \\ \hline \end{array}$
$\begin{array}{r} 4 \\ \times\ 5 \\ \hline \end{array}$
$\begin{array}{r} 3 \\ \times\ 2 \\ \hline \end{array}$

E.   3 x 9 = _____   4 x 3 = _____   4 x 7 = _____

F.   4 x 9 = _____   3 x 8 = _____   3 x 7 = _____

G.   3 x 6 = _____   4 x 8 = _____   4 x 6 = _____

41

Packaged Products

Multiply.

A. $4 \times 3 =$ __12__ $5 \times 4 =$ _____ $6 \times 4 =$ _____

B. $8 \times 3 =$ _____ $1 \times 4 =$ _____ $1 \times 3 =$ _____

C. $9 \times 4 =$ _____ $5 \times 3 =$ _____ $7 \times 3 =$ _____

D. $3 \times 3 =$ _____ $8 \times 4 =$ _____ $3 \times 4 =$ _____

E. $2 \times 3 =$ _____ $6 \times 3 =$ _____ $7 \times 4 =$ _____

F. $4 \times 4 =$ _____ $2 \times 4 =$ _____ $9 \times 3 =$ _____

G.
$$\begin{array}{r} 4 \\ \times\ 7 \\ \hline \end{array} \qquad \begin{array}{r} 9 \\ \times\ 3 \\ \hline \end{array} \qquad \begin{array}{r} 4 \\ \times\ 4 \\ \hline \end{array} \qquad \begin{array}{r} 3 \\ \times\ 5 \\ \hline \end{array} \qquad \begin{array}{r} 2 \\ \times\ 3 \\ \hline \end{array} \qquad \begin{array}{r} 3 \\ \times\ 7 \\ \hline \end{array}$$

H.
$$\begin{array}{r} 3 \\ \times\ 8 \\ \hline \end{array} \qquad \begin{array}{r} 1 \\ \times\ 4 \\ \hline \end{array} \qquad \begin{array}{r} 4 \\ \times\ 3 \\ \hline \end{array} \qquad \begin{array}{r} 7 \\ \times\ 3 \\ \hline \end{array} \qquad \begin{array}{r} 4 \\ \times\ 8 \\ \hline \end{array} \qquad \begin{array}{r} 3 \\ \times\ 3 \\ \hline \end{array}$$

I.
$$\begin{array}{r} 2 \\ \times\ 4 \\ \hline \end{array} \qquad \begin{array}{r} 3 \\ \times\ 6 \\ \hline \end{array} \qquad \begin{array}{r} 4 \\ \times\ 9 \\ \hline \end{array} \qquad \begin{array}{r} 8 \\ \times\ 4 \\ \hline \end{array} \qquad \begin{array}{r} 4 \\ \times\ 5 \\ \hline \end{array} \qquad \begin{array}{r} 3 \\ \times\ 2 \\ \hline \end{array}$$

J.
$$\begin{array}{r} 3 \\ \times\ 9 \\ \hline \end{array} \qquad \begin{array}{r} 4 \\ \times\ 2 \\ \hline \end{array} \qquad \begin{array}{r} 4 \\ \times\ 6 \\ \hline \end{array} \qquad \begin{array}{r} 3 \\ \times\ 1 \\ \hline \end{array}$$

                   FS-32070 Third Grade Math Review

Hop to It!

Multiply.

A. 4 x 1 = ___4___   6 x 0 = _____   1 x 9 = _____

B. 7 x 1 = _____   4 x 0 = _____   6 x 1 = _____

C. 9 x 0 = _____   1 x 1 = _____   1 x 7 = _____

D. 9 x 1 = _____   8 x 0 = _____   3 x 1 = _____

E. 0 x 7 = _____   2 x 1 = _____   8 x 1 = _____

F. 2 x 0 = _____   5 x 1 = _____   0 x 0 = _____

G.
```
    0        1        1        0        1        0
  x 4      x 8      x 2      x 5      x 0      x 8
```

H.
```
    1        3        1        0        1        5
  x 3      x 0      x 4      x 6      x 9      x 0
```

I.
```
    0        1        0        7        0        1
  x 9      x 6      x 3      x 0      x 2      x 5
```

FS-32070 Third Grade Math Review

Merry-Go-Round Multiplication

Multiply.

A.	3 x 0 ○	4 x 1	5 x 0	8 x 0	9 x 1	2 x 1
B.	6 x 0	3 x 1	9 x 0	7 x 1	8 x 1	1 x 0
C.	1 x 1	5 x 1	7 x 0	4 x 0	2 x 0	6 x 1

D. 0 x 1 = _____   0 x 6 = _____   1 x 2 = _____

E. 1 x 7 = _____   1 x 5 = _____   0 x 2 = _____

F. 0 x 4 = _____   0 x 5 = _____   1 x 3 = _____

G. 1 x 6 = _____   1 x 4 = _____   0 x 0 = _____

H. 1 x 8 = _____   0 x 3 = _____   1 x 9 = _____

I. 0 x 7 = _____   0 x 8 = _____   0 x 9 = _____

Crack the Code

Multiply. Fill in the correct letter over each answer.
Why is the letter B hot?

B ___ ___ ___ ___ ___ ___   ___ ___
14 49 18 35 54 56 49   28 42

___ ___ ___ ___ ___   ___ ___ ___   B ___ ___ ___
63 35 36 49 56   21 28 48   14 21 28 48

B 2 x 7 = 14	**C** 3 x 6	**F** 6 x 5	**G** 1 x 7	**A** 5 x 7	**D** 4 x 6
F 5 x 6	**H** 0 x 7	**I** 4 x 7	**J** 1 x 6	**M** 9 x 7	**O** 3 x 7
P 2 x 6	**K** 6 x 6	**S** 8 x 7	**T** 7 x 6	**H** 0 x 6	**A** 7 x 5
U 9 x 6	**T** 6 x 7	**O** 7 x 3	**E** 7 x 7	**D** 6 x 4	**L** 8 x 6

FS-32070 Third Grade Math Review

Play Ball!

Multiply.

A.
   5   4   6   2   1   7
x 6   x 7   x 7   x 6   x 7   x 6
30

B.
   4   3   6   1   5   9
x 6   x 7   x 6   x 6   x 7   x 6

C.
   2   3   8   7   8   9
x 7   x 6   x 6   x 7   x 7   x 7

Match each fact with its product.

D. 6 x 4   28   7 x 5   48

E. 7 x 4   12   6 x 8   18

F. 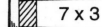 7 x 3   24   7 x 8   35

G. 6 x 2   21   6 x 3   14

H.  6 x 5   63   7 x 2   56

I. 7 x 9   54   6 x 7   6

J.  6 x 9   30   1 x 6   42

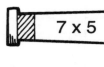
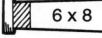

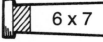
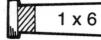

FS-32070 Third Grade Math Review

Diving Into Multiplication

Multiply.

A.
```
    4        2        4        7        6        8
  x 8      x 9      x 9      x 9      x 8      x 4
  ---
   32
```

B.
```
    8        3        8        8        5        8
  x 9      x 8      x 9      x 8      x 9      x 5
```

C.
```
    3        9        7        5        6        9
  x 9      x 8      x 8      x 8      x 9      x 9
```

Match each fact with its product.

D.   8 x 3 •   • 9

E.   9 x 5 •   • 24

F.   9 x 1 •   • 48

G.   8 x 6 •   • 18

H.   9 x 7 •   • 45

I.   9 x 2 •   • 56

J.   8 x 7 •   • 63

FS-32070 Third Grade Math Review

Name _____

Multiplication Bumper Cars

Multiply.

A.
  2   9   7   8   5   8
x 8   x 9   x 8   x 9   x 8   x 6
16

B.
  9   4   6   8   9   3
x 8   x 8   x 9   x 4   x 4   x 8

C.
  8   9   4   8   9   6
x 8   x 6   x 9   x 3   x 7   x 8

D.
  9   9   8   1   9   8
x 2   x 5   x 9   x 8   x 9   x 5

E.
  8   8   2   0   7   9
x 7   x 1   x 9   x 9   x 8   x 1

F.
  5   8   3   7   8   9
x 9   x 2   x 9   x 9   x 5   x 3

FS-32070 Third Grade Math Review

Facts and Figures

Complete the table.

X	0	1	2	3	4	5	6	7	8	9
0	0									
1									8	
2										
3										
4				12						
5										
6						30				
7			14					49		
8										
9										

FS-32070 Third Grade Math Review

Input–Output

Complete the tables.

A.

x	2
8	16
6	
5	
3	
9	
2	
7	
1	
4	
0	

B.

x	4
5	
2	
7	
9	
0	
6	
1	
3	
8	
4	

C.

x	7
6	
9	
1	
8	
3	
5	
0	
2	
4	
7	

D.

x	5
4	
2	
8	
3	
7	
9	
0	
6	
1	
5	

E.

x	3
7	
3	
6	
9	
2	
0	
8	
5	
1	
4	

F.

x	6
9	
3	
5	
2	
8	
6	
0	
7	
4	
1	

G.

x	8
3	
8	
4	
9	
0	
1	
7	
5	
2	
6	

H.

x	9
2	
8	
1	
4	
7	
3	
9	
6	
0	
5	

Same Size Sets

Circle sets of two. Divide.

A.

$4 \div 2 = \underline{\quad 2 \quad}$ $6 \div 2 = \underline{\qquad}$ $12 \div 2 = \underline{\qquad}$

B.

$8 \div 2 = \underline{\qquad}$ $4 \div 2 = \underline{\qquad}$ $10 \div 2 = \underline{\qquad}$

Circle sets of five. Divide.

C.

$5 \div 5 = \underline{\quad 1 \quad}$ $10 \div 5 = \underline{\qquad}$ $15 \div 5 = \underline{\qquad}$

D.

$25 \div 5 = \underline{\qquad}$ $35 \div 5 = \underline{\qquad}$ $20 \div 5 = \underline{\qquad}$

Divide.

E. $16 \div 2 = \underline{\quad 8 \quad}$  $2 \div 2 = \underline{\qquad}$ $30 \div 5 = \underline{\qquad}$

F. $40 \div 5 = \underline{\qquad}$ $45 \div 5 = \underline{\qquad}$ $4 \div 2 = \underline{\qquad}$

G. $10 \div 5 = \underline{\qquad}$ $8 \div 2 = \underline{\qquad}$ $35 \div 5 = \underline{\qquad}$

FS-32070 Third Grade Math Review

Winter Fun

Divide.

A.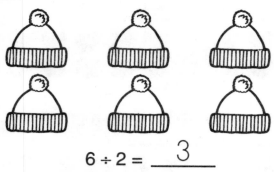

$6 \div 2 =$ ___3___ $10 \div 2 =$ _____

B.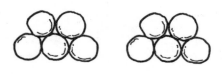

$10 \div 5 =$ _____ $20 \div 5 =$ _____

C.

$30 \div 5 =$ _____     $8 \div 2 =$ _____

Divide.

D. $14 \div 2 =$ ___7___ $4 \div 2 =$ _____ $12 \div 2 =$ _____

E. $15 \div 5 =$ _____ $25 \div 5 =$ _____ $2 \div 2 =$ _____

F. $5 \div 5 =$ _____ $16 \div 2 =$ _____ $18 \div 2 =$ _____

G. $45 \div 5 =$ _____ $40 \div 5 =$ _____ $35 \div 5 =$ _____

52   FS-32070 Third Grade Math Review

Name _____

Building Blocks

Divide.

A.

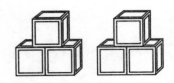

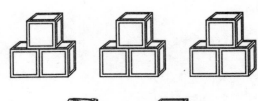

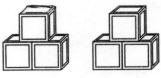

$9 \div 3 =$ ___3___ $15 \div 3 =$ _____

B.

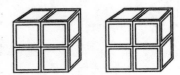

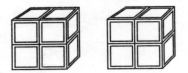

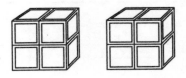

$16 \div 4 =$ _____ $20 \div 4 =$ _____

C. $6 \div 3 =$ ___2___ $4 \div 4 =$ _____ $24 \div 4 =$ _____

D. $12 \div 3 =$ _____ $21 \div 3 =$ _____ $12 \div 4 =$ _____

E. $3 \div 3 =$ _____ $32 \div 4 =$ _____ $16 \div 4 =$ _____

F. $28 \div 4 =$ _____ $24 \div 3 =$ _____ $15 \div 3 =$ _____

G. $8 \div 4 =$ _____ $6 \div 3 =$ _____ $20 \div 4 =$ _____

H. $27 \div 3 =$ _____ $36 \div 4 =$ _____ $18 \div 3 =$ _____

                                       FS-32070 Third Grade Math Review

Buttons and Bows

Divide.

A.

6 ÷ 3 = ___2___   9 ÷ 3 = _____   12 ÷ 3 = _____

B.

8 ÷ 4 = _____   12 ÷ 4 = _____   4 ÷ 4 = _____

C.   6 ÷ 3 = _____   15 ÷ 3 = _____   8 ÷ 4 = _____

D.   12 ÷ 3 = _____   24 ÷ 4 = _____   12 ÷ 4 = _____

E.   24 ÷ 3 = _____   18 ÷ 3 = _____   21 ÷ 3 = _____

F. $3\overline{)9}$ 3 $3\overline{)15}$ $4\overline{)8}$ $4\overline{)16}$ $3\overline{)21}$

G. $4\overline{)28}$ $4\overline{)20}$ $3\overline{)27}$ $4\overline{)24}$ $3\overline{)24}$

H. $3\overline{)18}$ $3\overline{)6}$ $4\overline{)32}$ $4\overline{)36}$ $3\overline{)3}$

54   FS-32070 Third Grade Math Review

Flower Power

$3 \div 1 = $ __3__ $3 \div 3 = $ __1__ $0 \div 3 = $ __0__

Divide.

A. $0 \div 1 = $ _____ $8 \div 1 = $ _____ $5 \div 5 = $ _____

B. $6 \div 1 = $ _____ $4 \div 4 = $ _____ $4 \div 1 = $ _____

C. $0 \div 5 = $ _____ $2 \div 1 = $ _____ $0 \div 6 = $ _____

D. $9 \div 9 = $ _____ $0 \div 3 = $ _____ $7 \div 7 = $ _____

E. $0 \div 4 = $ _____ $7 \div 1 = $ _____ $5 \div 1 = $ _____

F. $3 \div 3 = $ _____ $3 \div 1 = $ _____ $0 \div 8 = $ _____

G. $0 \div 9 = $ _____ $1 \div 1 = $ _____ $0 \div 7 = $ _____

H. $1 \div 1 = $ _____ $2 \div 2 = $ _____ $6 \div 6 = $ _____

I. $8 \div 8 = $ _____ $0 \div 2 = $ _____ $9 \div 1 = $ _____

Name _____

Cookie Capers

Divide.

A. $1\overline{)9}$ (answer: 9)
A. $1\overline{)9}$ $1\overline{)8}$ $1\overline{)7}$ $1\overline{)6}$ $1\overline{)5}$

B. $1\overline{)4}$ $1\overline{)3}$ $1\overline{)2}$ $1\overline{)1}$ $1\overline{)0}$

C. $9\overline{)9}$ $8\overline{)8}$ $7\overline{)7}$ $6\overline{)6}$ $5\overline{)5}$

D. $4\overline{)4}$ $3\overline{)3}$ $2\overline{)2}$ $1\overline{)1}$ $0\overline{)0}$

E. $9\overline{)0}$ $8\overline{)0}$ $7\overline{)0}$ $6\overline{)0}$

F. $5\overline{)0}$ $4\overline{)0}$ $3\overline{)0}$ $2\overline{)0}$

G. What is the quotient when a number is divided by itself?

_____

H. What is the quotient when a number is divided by 1?

_____

I. What is the quotient when 0 is divided by a number?

_____

Skiing Through Division

Divide.

A. $6\overline{)42}$ **7** $7\overline{)14}$ $6\overline{)18}$ $7\overline{)28}$ $6\overline{)48}$

B. $6\overline{)24}$ $7\overline{)21}$ $7\overline{)35}$ $6\overline{)54}$ $7\overline{)56}$

C. $7\overline{)42}$ $6\overline{)0}$ $6\overline{)12}$ $7\overline{)7}$ $7\overline{)49}$

D. $7\overline{)0}$ $7\overline{)63}$ $6\overline{)6}$ $6\overline{)30}$ $6\overline{)36}$

E. $48 \div 6 =$ ___**8**___ $56 \div 7 =$ _____ $12 \div 6 =$ _____

$6\overline{)48}$

F. $42 \div 6 =$ _____ $28 \div 7 =$ _____ $35 \div 7 =$ _____

G. $7 \div 7 =$ _____ $49 \div 7 =$ _____ $54 \div 6 =$ _____

H. $30 \div 6 =$ _____ $42 \div 7 =$ _____ $21 \div 7 =$ _____

I. $63 \div 7 =$ _____ $36 \div 6 =$ _____ $14 \div 7 =$ _____

          FS-32070 Third Grade Math Review

Name _____

Kangaroo Power

Divide.

A. 28 ÷ 7 = ___4___   14 ÷ 7 = _____   6 ÷ 1 = _____

B. 18 ÷ 6 = _____   42 ÷ 6 = _____   21 ÷ 7 = _____

C. 0 ÷ 7 = _____   56 ÷ 7 = _____   24 ÷ 6 = _____

D. 35 ÷ 7 = _____   54 ÷ 6 = _____   0 ÷ 6 = _____

E. 7 ÷ 7 = _____   12 ÷ 6 = _____   30 ÷ 6 = _____

F. 42 ÷ 7 = _____   36 ÷ 6 = _____   63 ÷ 7 = _____

G. 48 ÷ 6 = _____   49 ÷ 7 = _____   14 ÷ 7 = _____

H. $6\overline{)6}$ $6\overline{)12}$ $7\overline{)0}$ $7\overline{)42}$ $6\overline{)30}$

I. $7\overline{)21}$ $6\overline{)24}$ $7\overline{)35}$

J. $6\overline{)48}$ $7\overline{)63}$ $7\overline{)56}$

FS-32070 Third Grade Math Review

Skydiving Fun

Divide.

A. $8 \div 8 = $ _____ 1 $40 \div 8 = $ _____ $24 \div 8 = $ _____

B. $16 \div 8 = $ _____ $56 \div 8 = $ _____ $48 \div 8 = $ _____

C. $32 \div 8 = $ _____ $72 \div 8 = $ _____ $64 \div 8 = $ _____

D. $18 \div 9 = $ _____ $9 \div 9 = $ _____ $72 \div 9 = $ _____

E. $36 \div 9 = $ _____ $27 \div 9 = $ _____ $63 \div 9 = $ _____

F. $54 \div 9 = $ _____ $45 \div 9 = $ _____ $81 \div 9 = $ _____

G. $9 \overline{)0}$ ⁰ $9 \overline{)18}$ $9 \overline{)27}$

H. $8 \overline{)8}$ $9 \overline{)81}$ $8 \overline{)16}$

I. $8 \overline{)32}$ $9 \overline{)54}$ $9 \overline{)72}$

J. $8 \overline{)64}$ $8 \overline{)24}$ $9 \overline{)36}$ $8 \overline{)56}$ $8 \overline{)48}$

59

Name _____

Gifts

Choose a number from the gift box to complete each division sentence.

A. $24 \div 8 = \boxed{3}$

B. $48 \div 8 = \boxed{}$

C. $72 \div 8 = \boxed{}$

$56 \div 8 = \boxed{}$

$16 \div 8 = \boxed{}$

$40 \div 8 = \boxed{}$

D. $\boxed{} \div 8 = 1$

E. $\boxed{} \div 8 = 2$

F. $\boxed{} \div 8 = 3$

$\boxed{} \div 8 = 8$

$\boxed{} \div 8 = 6$

$\boxed{} \div 8 = 4$

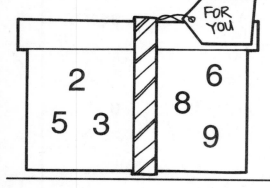

G. $27 \div 9 = \boxed{}$

H. $81 \div 9 = \boxed{}$

I. $54 \div 9 = \boxed{}$

$18 \div 9 = \boxed{}$

$72 \div 9 = \boxed{}$

$45 \div 9 = \boxed{}$

J. $\boxed{} \div 9 = 1$

K. $\boxed{} \div 9 = 7$

L. $\boxed{} \div 9 = 4$

$\boxed{} \div 9 = 8$

$\boxed{} \div 9 = 6$

$\boxed{} \div 9 = 5$

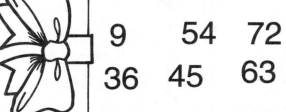

FS-32070 Third Grade Math Review

Name _____

Fishing for Quotients

Divide. Color the fish.

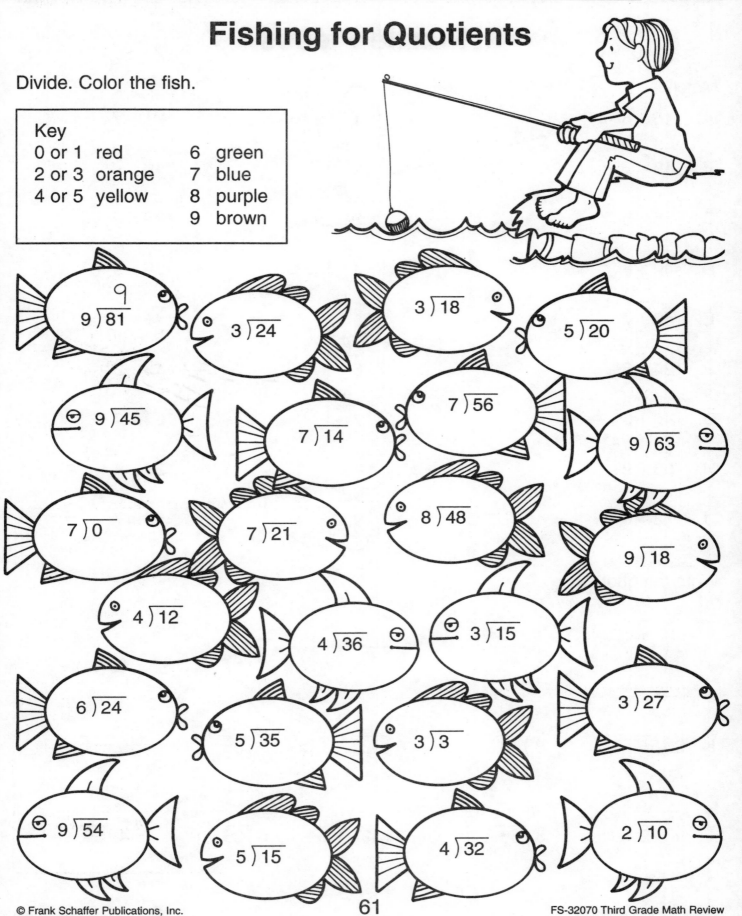

Key

0 or 1	red	
2 or 3	orange	6 green
4 or 5	yellow	7 blue
		8 purple
		9 brown

$9\overline{)81}$ 9

$3\overline{)24}$

$3\overline{)18}$

$5\overline{)20}$

$9\overline{)45}$

$7\overline{)14}$

$7\overline{)56}$

$9\overline{)63}$

$7\overline{)0}$

$7\overline{)21}$

$8\overline{)48}$

$9\overline{)18}$

$4\overline{)12}$

$4\overline{)36}$

$3\overline{)15}$

$6\overline{)24}$

$5\overline{)35}$

$3\overline{)3}$

$3\overline{)27}$

$9\overline{)54}$

$5\overline{)15}$

$4\overline{)32}$

$2\overline{)10}$

FS-32070 Third Grade Math Review

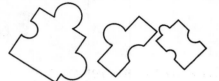

Puzzle Pieces

Match.

A. 16 ÷ 8

B. 18 ÷ 6

C. 5 ÷ 5

D. 40 ÷ 8

E. 42 ÷ 7

F. 24 ÷ 6

G. 40 ÷ 5

H. 63 ÷ 9

I. 72 ÷ 8

1

2

3

4

5

6

7

8

9

21 ÷ 7

10 ÷ 5

9 ÷ 9

36 ÷ 9

45 ÷ 9

18 ÷ 3

63 ÷ 7

28 ÷ 4

56 ÷ 7

Write the quotient.

J. $6\overline{)42}$ (7) $7\overline{)28}$ $5\overline{)30}$ $8\overline{)24}$ $7\overline{)21}$

K. $9\overline{)54}$ $8\overline{)48}$ $6\overline{)48}$ $7\overline{)35}$ $9\overline{)81}$

L. $9\overline{)72}$ $8\overline{)56}$ $7\overline{)49}$ $6\overline{)36}$ $4\overline{)36}$

FS-32070 Third Grade Math Review

Name _____

Family Albums

Multiply or divide.
Write the numbers for each fact family in the album label.

A.

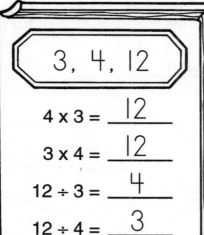

3, 4, 12

4 x 3 = __12__

3 x 4 = __12__

12 ÷ 3 = __4__

12 ÷ 4 = __3__

5 x 4 = _____

4 x 5 = _____

20 ÷ 5 = _____

20 ÷ 4 = _____

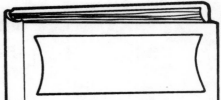

2 x 9 = _____

9 x 2 = _____

18 ÷ 2 = _____

18 ÷ 9 = _____

B.

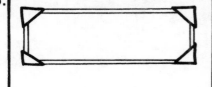

5 x 8 = _____

8 x 5 = _____

40 ÷ 5 = _____

40 ÷ 8 = _____

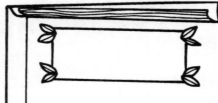

7 x 6 = _____

6 x 7 = _____

42 ÷ 7 = _____

42 ÷ 6 = _____

4 x 6 = _____

6 x 4 = _____

24 ÷ 4 = _____

24 ÷ 6 = _____

C.

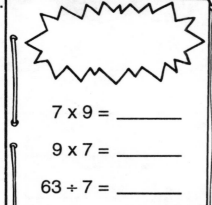

7 x 9 = _____

9 x 7 = _____

63 ÷ 7 = _____

63 ÷ 9 = _____

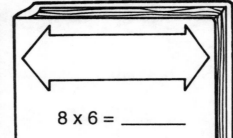

8 x 6 = _____

6 x 8 = _____

48 ÷ 6 = _____

48 ÷ 8 = _____

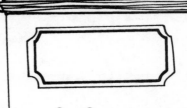

9 x 8 = _____

8 x 9 = _____

72 ÷ 9 = _____

72 ÷ 8 = _____

FS-32070 Third Grade Math Review

Team Jerseys

Write a fact family for each group of numbers.

A.

$3 \times 5 = 15$

$5 \times 3 = 15$

$15 \div 3 = 5$

$15 \div 5 = 3$

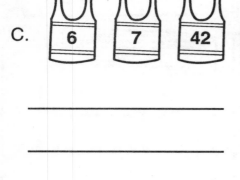

B.

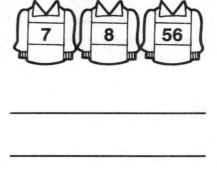

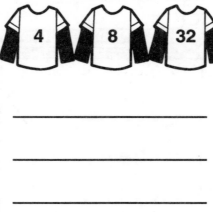

C.

FS-32070 Third Grade Math Review

Fraction Fun

Write a fraction for the part that is shaded and for the part that is not shaded.

A.

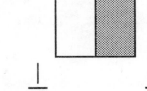

$\dfrac{1}{2}$ ____ shaded $\dfrac{1}{2}$ ____ not shaded

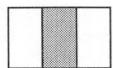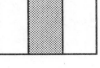

____ shaded ____ not shaded

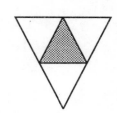

____ shaded ____ not shaded

B.

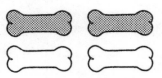

____ shaded ____ not shaded

____ shaded ____ not shaded

____ shaded ____ not shaded

C.

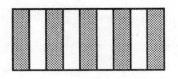

____ shaded ____ not shaded

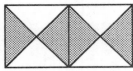

____ shaded ____ not shaded

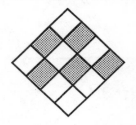

____ shaded ____ not shaded

D.

____ shaded ____ not shaded

____ shaded ____ not shaded

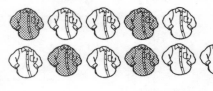

____ shaded ____ not shaded

FS-32070 Third Grade Math Review

Name_____ Fractions

Made in the Shade

Write a fraction for the part that is shaded and for the part that is not shaded.

A.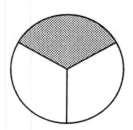

$$\frac{1}{3}$$ _____ shaded

$$\frac{2}{3}$$ _____ not shaded

B.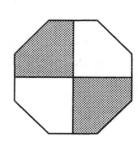

_____ shaded

_____ not shaded

C.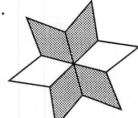

_____ shaded

_____ not shaded

D.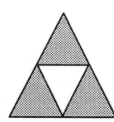

_____ shaded

_____ not shaded

E.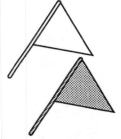

_____ shaded

_____ not shaded

F.

_____ shaded

_____ not shaded

G.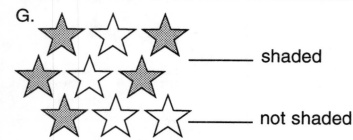

_____ shaded

_____ not shaded

H.

_____ shaded

_____ not shaded

I.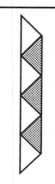

_____ shaded

_____ not shaded

J.

_____ shaded

_____ not shaded

FS-32070 Third Grade Math Review

A Piece of the Pie

Write <, >, or = in the 〇

A.
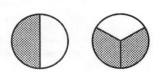

$\frac{1}{2}$ 〈<〉 $\frac{2}{3}$

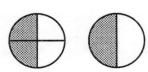

$\frac{2}{4}$ 〇 $\frac{1}{2}$

$\frac{1}{5}$ 〇 $\frac{1}{10}$

B.
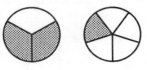

$\frac{2}{3}$ 〇 $\frac{1}{5}$

$\frac{1}{4}$ 〇 $\frac{1}{5}$

$\frac{1}{2}$ 〇 $\frac{5}{10}$

C.
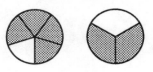

$\frac{4}{5}$ 〇 $\frac{2}{3}$

$\frac{1}{3}$ 〇 $\frac{2}{3}$

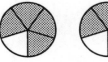

$\frac{4}{5}$ 〇 $\frac{2}{5}$

D.
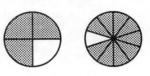

$\frac{3}{4}$ 〇 $\frac{9}{10}$

$\frac{2}{4}$ 〇 $\frac{3}{5}$

$\frac{1}{3}$ 〇 $\frac{1}{2}$

E.

$\frac{2}{5}$ 〇 $\frac{4}{10}$

$\frac{2}{3}$ 〇 $\frac{3}{10}$

$\frac{9}{10}$ 〇 $\frac{1}{2}$

 FS-32070 Third Grade Math Review

Name _____

Compare and Share

Write <, >, or = in the ◯.
Use the fraction bars to help you.

$\frac{1}{2}$					$\frac{1}{2}$				
$\frac{1}{3}$		$\frac{1}{3}$			$\frac{1}{3}$				
$\frac{1}{4}$		$\frac{1}{4}$		$\frac{1}{4}$		$\frac{1}{4}$			
$\frac{1}{5}$		$\frac{1}{5}$		$\frac{1}{5}$		$\frac{1}{5}$		$\frac{1}{5}$	
$\frac{1}{8}$	$\frac{1}{8}$	$\frac{1}{8}$	$\frac{1}{8}$	$\frac{1}{8}$	$\frac{1}{8}$	$\frac{1}{8}$	$\frac{1}{8}$		
$\frac{1}{10}$	$\frac{1}{10}$	$\frac{1}{10}$	$\frac{1}{10}$	$\frac{1}{10}$	$\frac{1}{10}$	$\frac{1}{10}$	$\frac{1}{10}$	$\frac{1}{10}$	$\frac{1}{10}$

A. $\frac{1}{2}$ ⓥ $\frac{1}{3}$ $\qquad$ $\frac{1}{2}$ ◯ $\frac{2}{4}$ $\qquad$ $\frac{1}{5}$ ◯ $\frac{1}{4}$

B. $\frac{3}{10}$ ◯ $\frac{1}{4}$ $\qquad$ $\frac{1}{3}$ ◯ $\frac{1}{10}$ $\qquad$ $\frac{3}{5}$ ◯ $\frac{6}{10}$

C. $\frac{3}{8}$ ◯ $\frac{3}{4}$ $\qquad$ $\frac{1}{2}$ ◯ $\frac{2}{5}$ $\qquad$ $\frac{3}{5}$ ◯ $\frac{2}{3}$

D. $\frac{2}{4}$ ◯ $\frac{4}{8}$ $\qquad$ $\frac{1}{3}$ ◯ $\frac{1}{5}$ $\qquad$ $\frac{4}{5}$ ◯ $\frac{8}{10}$

E. $\frac{1}{5}$ ◯ $\frac{1}{10}$ $\qquad$ $\frac{1}{4}$ ◯ $\frac{1}{8}$ $\qquad$ $\frac{1}{5}$ ◯ $\frac{3}{10}$

Write the fractions in order from least to greatest.

F. $\frac{2}{5}, \frac{3}{4}, \frac{1}{2}$ _____ $\qquad$ $\frac{1}{8}, \frac{4}{5}, \frac{9}{10}$ _____

G. $\frac{3}{4}, \frac{3}{10}, \frac{3}{8}$ _____ $\qquad$ $\frac{2}{5}, \frac{6}{8}, \frac{5}{10}$ _____

H. $\frac{1}{2}, \frac{5}{8}, \frac{2}{3}$ _____ $\qquad$ $\frac{4}{5}, \frac{1}{4}, \frac{2}{10}$ _____

68

Mix and Match

Write a mixed number for the part that is shaded.

A.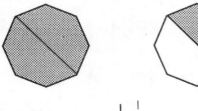

$1\frac{1}{2}$ _____ _____

B.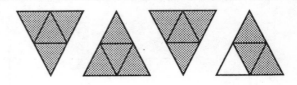

_____ _____

C.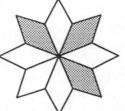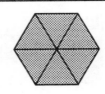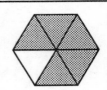

_____ _____

Match.

D. $3\frac{1}{4}$

E. 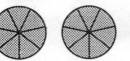 $2\frac{6}{7}$

F. 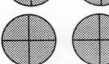 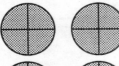 $3\frac{1}{8}$

G. $2\frac{4}{5}$

Mixed Numbers

Color to show the mixed number.

A. $1\frac{2}{3}$ $2\frac{1}{2}$

B. $3\frac{1}{4}$ $2\frac{5}{8}$

C.    $1\frac{5}{6}$  $1\frac{7}{10}$

 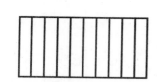

Write a mixed number for the shaded part.

D.

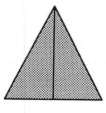   _____   _____

E.

 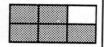     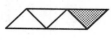

_____   _____

70

Name It!

Complete the chart.

	Picture	Mixed Number	Decimal	Word
A.		$1\frac{3}{10}$	1.3	__1__ and __3__ tenths
B.				_____ and _____ tenths
C.				_____ and _____ tenth
D.				_____ and _____ tenths
E.				_____ and _____ tenths
F.				_____ and _____ tenths
G.				_____ and _____ tenths

FS-32070 Third Grade Math Review

See and Say

Match.

A. 1.8

B. 1.3

C. 2.4

D. 1.2

E. 1.7

F. 2.3

G. 2.6

H. 2.5

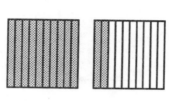

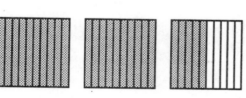
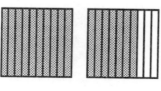
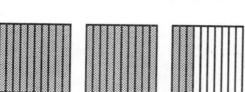

one and two tenths

one and three tenths

two and four tenths

two and six tenths

one and eight tenths

one and seven tenths

two and three tenths

two and five tenths

Name _____

Size It Up

Compare. Write > or < in the ◯.

A. 1.1 ⊙< 1.3   3.5 ◯ 2.7   4.5 ◯ 4.7

B. 4.9 ◯ 5.3   2.7 ◯ 1.9   3.5 ◯ 2.8

C. 2.2 ◯ 1.1   5.6 ◯ 5.7   9.2 ◯ 8.9

D. 8.0 ◯ 7.8   2.6 ◯ 6.2   1.9 ◯ 2.0

E. 2.4 ◯ 2.0   3.9 ◯ 2.5   1.8 ◯ 1.3

F. 3.5 ◯ 3.1   4.2 ◯ 3.8   7.1 ◯ 7.2

G. 6.2 ◯ 7.0   8.3 ◯ 8.9   9.5 ◯ 9.4

Write the decimals from least to greatest.

H. 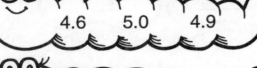 2.3   1.9   3.7   1.9, 2.3, 3.7

I.   4.6   5.0   4.9   _____

J.   3.8   4.1   2.9   _____

K.  8.2   7.9   8.1   _____

FS-32070 Third Grade Math Review

Name_____

Leaping Along

Compare. Write < or > in the ◯ .

A.   3.7 ⟨>⟩ 3.4   2.5 ◯ 2.9   4.2 ◯ 5.2

B.   6.1 ◯ 6.8   4.6 ◯ 3.6   8.9 ◯ 9.5

C.   2.8 ◯ 3.1   1.7 ◯ 1.1   4.3 ◯ 4.0

D.   3.5 ◯ 3.6   9.8 ◯ 8.9   7.3 ◯ 6.8

E.   3.1 ◯ 3.8   1.4 ◯ 2.6   3.5 ◯ 3.9

F.   4.8 ◯ 5.1   6.2 ◯ 6.5   5.7 ◯ 5.8

G.   6.3 ◯ 5.8   7.2 ◯ 7.3   8.5 ◯ 9.2

H.   8.9 ◯ 9.3   7.4 ◯ 6.9   4.6 ◯ 5.3

Write the decimals from least to greatest.

I.   2.3,   2.0,   2.4    2.0   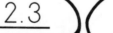 2.3   2.4

J.   6.5,   6.2,   6.7   ____   ____   ____

K.   5.0,   5.1,   4.9   ____   ____   ____

L.   7.8,   7.9,   7.7   ____   ____   ____

FS-32070 Third Grade Math Review

Decimal Sums and Differences

Add.

A. 3.4 + 2.5 5.9	2.6 + 3.7	2.5 + 5.5	1.9 + 6.6	1.0 + 8.8

B. 2.7 + 6.3	3.7 + 5.4	2.9 + 3.1	8.6 + 1.5	2.5 + 1.7

Subtract.

C. 5.9 − 2.5	1.3 − 0.9	8.2 − 6.3	7.4 − 2.8	6.5 − 1.6

D. 4.7 − 1.3	8.6 − 4.2	9.3 − 7.6	8.0 − 6.8	7.6 − 5.6

Write the decimal and solve.

E. Two and seven tenths plus three and nine tenths.

F. Six and eight tenths minus four and three tenths.

Adding and Subtracting Decimals

Write the decimal and solve.

A. Two and three tenths plus five and nine tenths.

B. Six and eight tenths plus one and seven tenths.

C. Six and five tenths minus four and two tenths.

D. Three and two tenths minus two and three tenths.

Add.

E.	4.4	3.5	6.5	9.3	6.7
	+ 1.7	+ 2.9	+ 1.3	+ 0.5	+ 2.9

F.	8.4	7.8	4.3	8.5	3.6
	+ 1.3	+ 1.9	+ 2.7	+ 1.0	+ 4.6

Subtract.

G.	9.8	4.7	8.1	4.3	9.3
	− 5.2	− 2.9	− 5.1	− 3.9	− 6.8

H.	6.0	7.5	5.3	8.6	7.2
	− 1.5	− 5.0	− 2.7	− 8.0	− 4.8

FS-32070 Third Grade Math Review

Time for Fun

Draw a line from each clock to the clock with the matching time.

A.

F.

| 3:48 |

| 8:21 |

B.

G.

| 12:38 |

| 1:57 |

C.

H.

| 9:08 |

| 2:19 |

D.

I.

| 11:43 |

| 5:11 |

E.

J.

| 4:52 |

| 7:26 |

FS-32070 Third Grade Math Review

Name _____

Party Time

Write the time.

A.

_____ 1:12 _____   _____   _____

B.

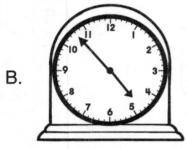   

_____   _____   _____   _____

C.

_____   _____   _____   _____

D.

_____   _____   _____   _____

FS-32070 Third Grade Math Review

Timers

Write how much time has passed.

A.  to

__1 hour 25 minutes__

B.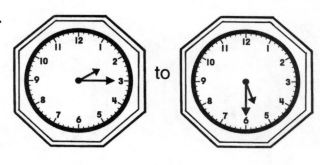

_____

C.  to

_____

D.  to

_____

E.  to

_____

F.  to

_____

G.  to

_____

H.  to

_____

FS-32070 Third Grade Math Review

How Long Will It Take?

Write how much time has passed.

A.
 to

2 hours 15 minutes

B.
 to

C.
 to

D.
 to

E.
 to

F.
 to

G.
 to

H.
 to

What's the Change?

Cross out the change. Write the amount.

	Item Bought	Money Paid	Coins for Change	Amount of Change
A.	$0.39			61¢
B.	$1.27			_____
C.	$2.98			_____
D.	$0.85			_____
E.	$1.07			_____
F.	$1.53			_____

FS-32070 Third Grade Math Review

Grocery Shopping

Mark the coins you would get for change. Use the fewest coins you can. Write the amount of change.

	Item Bought	Money Paid	Your Change						Amount of Change
A.	apple $0.34	$0.50				I	I	I	16¢
B.	bread $2.44	$3.00							_____
C.	pretzel $0.76	$1.00							_____
D.	grapes $1.69	$2.00							_____
E.	Soup $1.55	$2.00							_____
F.	$3.98	$5.00							_____

Shopping Spree

Add.

A. $\begin{array}{r} \overset{1\ 1}{\$3.25} \\ +2.87 \\ \hline \$6.12 \end{array}$
$\begin{array}{r} \$2.83 \\ +1.97 \\ \hline \end{array}$
$\begin{array}{r} \$4.56 \\ +1.28 \\ \hline \end{array}$
$\begin{array}{r} \$6.80 \\ +1.49 \\ \hline \end{array}$
$\begin{array}{r} \$3.27 \\ +2.58 \\ \hline \end{array}$

B. $\begin{array}{r} \$4.65 \\ +1.83 \\ \hline \end{array}$
$\begin{array}{r} \$3.79 \\ +4.44 \\ \hline \end{array}$
$\begin{array}{r} \$2.15 \\ +3.97 \\ \hline \end{array}$
$\begin{array}{r} \$5.63 \\ +3.79 \\ \hline \end{array}$
$\begin{array}{r} \$7.18 \\ +2.76 \\ \hline \end{array}$

Subtract.

C. $\begin{array}{r} \overset{8\ 18}{\$2.98} \\ -1.49 \\ \hline \$1.49 \end{array}$
$\begin{array}{r} \$3.75 \\ -2.90 \\ \hline \end{array}$
$\begin{array}{r} \$4.02 \\ -3.98 \\ \hline \end{array}$
$\begin{array}{r} \$8.59 \\ -2.75 \\ \hline \end{array}$
$\begin{array}{r} \$7.94 \\ -5.47 \\ \hline \end{array}$

D. $\begin{array}{r} \$6.73 \\ -2.85 \\ \hline \end{array}$
$\begin{array}{r} \$5.84 \\ -2.95 \\ \hline \end{array}$
$\begin{array}{r} \$7.00 \\ -3.53 \\ \hline \end{array}$
$\begin{array}{r} \$9.05 \\ -3.49 \\ \hline \end{array}$
$\begin{array}{r} \$7.37 \\ -3.64 \\ \hline \end{array}$

Write a number sentence and solve.

E. Serene bought a cap for $5.15 and a book for $2.95. How much did she spend in all?

F. Paul had $5.00. He spent $2.97 on stickers. How much money does he have left?

83

Name _____

Money in the Bank

Add or subtract.

A. $2.29 $4.53 $5.51 $6.84 $3.69
   +6.87   −3.98   +2.99   −3.49   +4.73
   $9.16

B. $2.86 $5.92 $4.95 $1.57 $8.48
   +4.25   −1.85   −4.07   +4.98   −3.72

C. $9.47 $4.02 $3.42 $8.35 $6.09
   −3.89   −1.95   +4.57   −3.94   +5.25

D. $8.00 $2.57 $8.42 $2.68 $7.32
   −4.49   +3.98   −1.79   +3.99   −3.21

E. $2.48 $7.03 $5.15 $2.89 $6.45
   +4.29   −3.97   −2.84   +3.47   −4.32

Write a number sentence and solve.

F. Marc had $5.95 in his bank. Then he added $2.15 to it. How much is in Marc's bank now?

G. Rosie had $9.00. She spent $2.49 on a toy boat. How much does Rosie have now?

FS-32070 Third Grade Math Review

Name _____

Special Tens

Multiply.

A.
$$\begin{array}{r} 80 \\ \times\ 6 \\ \hline 480 \end{array}$$
$$\begin{array}{r} 30 \\ \times\ 5 \\ \hline \end{array}$$
$$\begin{array}{r} 60 \\ \times\ 2 \\ \hline \end{array}$$
$$\begin{array}{r} 50 \\ \times\ 4 \\ \hline \end{array}$$
$$\begin{array}{r} 40 \\ \times\ 6 \\ \hline \end{array}$$

B.
$$\begin{array}{r} 40 \\ \times\ 4 \\ \hline \end{array}$$
$$\begin{array}{r} 50 \\ \times\ 7 \\ \hline \end{array}$$
$$\begin{array}{r} 20 \\ \times\ 3 \\ \hline \end{array}$$
$$\begin{array}{r} 70 \\ \times\ 3 \\ \hline \end{array}$$
$$\begin{array}{r} 10 \\ \times\ 8 \\ \hline \end{array}$$

C.
$$\begin{array}{r} 80 \\ \times\ 4 \\ \hline \end{array}$$
$$\begin{array}{r} 90 \\ \times\ 4 \\ \hline \end{array}$$
$$\begin{array}{r} 10 \\ \times\ 5 \\ \hline \end{array}$$
$$\begin{array}{r} 20 \\ \times\ 7 \\ \hline \end{array}$$
$$\begin{array}{r} 30 \\ \times\ 9 \\ \hline \end{array}$$

D.
$$\begin{array}{r} 40 \\ \times\ 7 \\ \hline \end{array}$$
$$\begin{array}{r} 50 \\ \times\ 6 \\ \hline \end{array}$$
$$\begin{array}{r} 70 \\ \times\ 8 \\ \hline \end{array}$$
$$\begin{array}{r} 60 \\ \times\ 6 \\ \hline \end{array}$$
$$\begin{array}{r} 70 \\ \times\ 2 \\ \hline \end{array}$$

E.
$$\begin{array}{r} 90 \\ \times\ 6 \\ \hline \end{array}$$
$$\begin{array}{r} 80 \\ \times\ 8 \\ \hline \end{array}$$
$$\begin{array}{r} 30 \\ \times\ 8 \\ \hline \end{array}$$
$$\begin{array}{r} 40 \\ \times\ 5 \\ \hline \end{array}$$
$$\begin{array}{r} 60 \\ \times\ 8 \\ \hline \end{array}$$

F.
$$\begin{array}{r} 70 \\ \times\ 7 \\ \hline \end{array}$$
$$\begin{array}{r} 90 \\ \times\ 7 \\ \hline \end{array}$$
$$\begin{array}{r} 60 \\ \times\ 9 \\ \hline \end{array}$$
$$\begin{array}{r} 30 \\ \times\ 7 \\ \hline \end{array}$$
$$\begin{array}{r} 90 \\ \times\ 9 \\ \hline \end{array}$$

FS-32070 Third Grade Math Review

Name _____

 **Wonderful Hundreds**

Multiply.

A.   5 x 1 = _____   7 x 7 = _____   3 x 5 = _____

        5 x 10 = _____   7 x 70 = _____   3 x 50 = _____

        5 x 100 = _____   7 x 700 = _____   3 x 500 = _____

B.   800   400   900   500   300
      x   2   x   8   x   6   x   2   x   3
      1,600

C.   600   200   700   300   400
      x   3   x   4   x   3   x   4   x   7

D.   800   500   100   600   900
      x   3   x   4   x   5   x   4   x   3

E.   200   400   300   900   500
      x   7   x   4   x   6   x   9   x   5

F.   700   500   600   100   200
      x   4   x   7   x   8   x   8   x   2

86   FS-32070 Third Grade Math Review

Strawberry Patch

Multiply.

A. 33
   x 2
   66

32
x 3

23
x 3

44
x 2

34
x 2

B.

21
x 4

12
x 2

10
x 5

22
x 3

41
x 2

C. 20
   x 4

14
x 2

23
x 2

43
x 2

11
x 3

D.

41
x 2

33
x 3

31
x 3

42
x 2

32
x 2

E. 22
   x 2

13
x 3

12
x 3

11
x 9

12
x 4

FS-32070 Third Grade Math Review

Name_____

Multiplication Paint

Multiply.

A.
$$\begin{array}{r} 12 \\ \times\ 4 \\ \hline 48 \end{array}$$

13
x 3

31
x 3

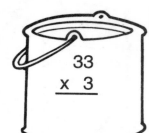
33
x 3

43
x 2

B.
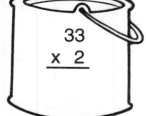
33
x 2

10
x 5

22
x 4

23
x 2

11
x 4

C.
41
x 2

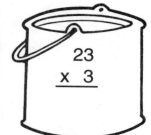
23
x 3

40
x 2

20
x 3

14
x 2

D.
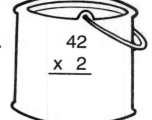
42
x 2

30
x 2

21
x 3

11
x 3

12
x 2

E.
20
x 4

31
x 2

30
x 3

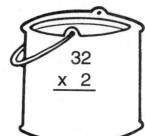
32
x 2

22
x 2

FS-32070 Third Grade Math Review

Name _____

A Buried Treasure

Multiply. Color the boxes whose products are greater than 70.

A.	$49 \times 2 = 98$	14×3	16×5	27×3	17×5	16×3
B.	45×2	19×5	23×4	17×2	29×3	15×3
C.	19×2	13×4	26×2	16×6	12×8	25×2
D.	14×6	38×2	26×3	13×6	27×2	15×4
E.	18×5	28×2	19×3	18×3	17×4	12×5
F.	14×7	48×2	13×5	24×4	28×3	
G.	18×3	13×7	12×7	46×2	17×3	

89

Name _____

Hooked on Multiplication

Multiply.

A.
$$\begin{array}{r} \overset{1}{2}3 \\ \times\ 4 \\ \hline 92 \end{array}$$

$$\begin{array}{r} 12 \\ \times\ 8 \\ \hline \end{array}$$
$$\begin{array}{r} 27 \\ \times\ 3 \\ \hline \end{array}$$
$$\begin{array}{r} 13 \\ \times\ 5 \\ \hline \end{array}$$
$$\begin{array}{r} 13 \\ \times\ 7 \\ \hline \end{array}$$

B.
$$\begin{array}{r} 47 \\ \times\ 2 \\ \hline \end{array}$$
$$\begin{array}{r} 16 \\ \times\ 3 \\ \hline \end{array}$$
$$\begin{array}{r} 45 \\ \times\ 2 \\ \hline \end{array}$$
$$\begin{array}{r} 37 \\ \times\ 2 \\ \hline \end{array}$$
$$\begin{array}{r} 16 \\ \times\ 5 \\ \hline \end{array}$$

C.
$$\begin{array}{r} 14 \\ \times\ 5 \\ \hline \end{array}$$
$$\begin{array}{r} 14 \\ \times\ 3 \\ \hline \end{array}$$
$$\begin{array}{r} 36 \\ \times\ 2 \\ \hline \end{array}$$
$$\begin{array}{r} 12 \\ \times\ 5 \\ \hline \end{array}$$
$$\begin{array}{r} 13 \\ \times\ 6 \\ \hline \end{array}$$

D.
$$\begin{array}{r} 46 \\ \times\ 2 \\ \hline \end{array}$$
$$\begin{array}{r} 25 \\ \times\ 3 \\ \hline \end{array}$$
$$\begin{array}{r} 19 \\ \times\ 3 \\ \hline \end{array}$$
$$\begin{array}{r} 49 \\ \times\ 2 \\ \hline \end{array}$$
$$\begin{array}{r} 24 \\ \times\ 4 \\ \hline \end{array}$$

E.
$$\begin{array}{r} 19 \\ \times\ 2 \\ \hline \end{array}$$
$$\begin{array}{r} 35 \\ \times\ 2 \\ \hline \end{array}$$
$$\begin{array}{r} 27 \\ \times\ 2 \\ \hline \end{array}$$
$$\begin{array}{r} 18 \\ \times\ 4 \\ \hline \end{array}$$
$$\begin{array}{r} 12 \\ \times\ 6 \\ \hline \end{array}$$

F.
$$\begin{array}{r} 16 \\ \times\ 6 \\ \hline \end{array}$$
$$\begin{array}{r} 14 \\ \times\ 6 \\ \hline \end{array}$$
$$\begin{array}{r} 28 \\ \times\ 3 \\ \hline \end{array}$$
$$\begin{array}{r} 19 \\ \times\ 4 \\ \hline \end{array}$$
$$\begin{array}{r} 49 \\ \times\ 2 \\ \hline \end{array}$$

FS-32070 Third Grade Math Review

Look Out for Multiplication

Multiply.

A.
$$\begin{array}{r} \overset{3}{47} \\ \times\ 5 \\ \hline 235 \end{array}$$
$$\begin{array}{r} 23 \\ \times\ 7 \\ \hline \end{array}$$
$$\begin{array}{r} 17 \\ \times\ 9 \\ \hline \end{array}$$
$$\begin{array}{r} 95 \\ \times\ 2 \\ \hline \end{array}$$
$$\begin{array}{r} 27 \\ \times\ 3 \\ \hline \end{array}$$

B.
$$\begin{array}{r} 56 \\ \times\ 8 \\ \hline \end{array}$$
$$\begin{array}{r} 45 \\ \times\ 4 \\ \hline \end{array}$$
$$\begin{array}{r} 42 \\ \times\ 8 \\ \hline \end{array}$$
$$\begin{array}{r} 36 \\ \times\ 4 \\ \hline \end{array}$$
$$\begin{array}{r} 53 \\ \times\ 5 \\ \hline \end{array}$$

C.
$$\begin{array}{r} 44 \\ \times\ 7 \\ \hline \end{array}$$
$$\begin{array}{r} 27 \\ \times\ 5 \\ \hline \end{array}$$
$$\begin{array}{r} 59 \\ \times\ 9 \\ \hline \end{array}$$
$$\begin{array}{r} 83 \\ \times\ 6 \\ \hline \end{array}$$
$$\begin{array}{r} 37 \\ \times\ 8 \\ \hline \end{array}$$

D.
$$\begin{array}{r} 69 \\ \times\ 3 \\ \hline \end{array}$$
$$\begin{array}{r} 46 \\ \times\ 4 \\ \hline \end{array}$$
$$\begin{array}{r} 67 \\ \times\ 5 \\ \hline \end{array}$$
$$\begin{array}{r} 84 \\ \times\ 3 \\ \hline \end{array}$$
$$\begin{array}{r} 65 \\ \times\ 4 \\ \hline \end{array}$$

E.
$$\begin{array}{r} 75 \\ \times\ 3 \\ \hline \end{array}$$
$$\begin{array}{r} 19 \\ \times\ 9 \\ \hline \end{array}$$
$$\begin{array}{r} 45 \\ \times\ 7 \\ \hline \end{array}$$
$$\begin{array}{r} 26 \\ \times\ 8 \\ \hline \end{array}$$
$$\begin{array}{r} 99 \\ \times\ 3 \\ \hline \end{array}$$

F.
$$\begin{array}{r} 63 \\ \times\ 7 \\ \hline \end{array}$$
$$\begin{array}{r} 84 \\ \times\ 5 \\ \hline \end{array}$$
$$\begin{array}{r} 35 \\ \times\ 7 \\ \hline \end{array}$$

FS-32070 Third Grade Math Review

Croaking About Multiplication

Multiply.

A.
$$\begin{array}{r} \overset{4}{37} \\ \times\ 6 \\ \hline 222 \end{array}$$
$$\begin{array}{r} 27 \\ \times\ 4 \\ \hline \end{array}$$
$$\begin{array}{r} 46 \\ \times\ 5 \\ \hline \end{array}$$
$$\begin{array}{r} 35 \\ \times\ 7 \\ \hline \end{array}$$
$$\begin{array}{r} 56 \\ \times\ 4 \\ \hline \end{array}$$

B.
$$\begin{array}{r} 84 \\ \times\ 8 \\ \hline \end{array}$$
$$\begin{array}{r} 19 \\ \times\ 9 \\ \hline \end{array}$$
$$\begin{array}{r} 28 \\ \times\ 6 \\ \hline \end{array}$$
$$\begin{array}{r} 39 \\ \times\ 7 \\ \hline \end{array}$$
$$\begin{array}{r} 67 \\ \times\ 4 \\ \hline \end{array}$$

C.
$$\begin{array}{r} 48 \\ \times\ 5 \\ \hline \end{array}$$
$$\begin{array}{r} 17 \\ \times\ 9 \\ \hline \end{array}$$
$$\begin{array}{r} 38 \\ \times\ 7 \\ \hline \end{array}$$
$$\begin{array}{r} 65 \\ \times\ 4 \\ \hline \end{array}$$
$$\begin{array}{r} 59 \\ \times\ 5 \\ \hline \end{array}$$

D.
$$\begin{array}{r} 68 \\ \times\ 2 \\ \hline \end{array}$$
$$\begin{array}{r} 79 \\ \times\ 3 \\ \hline \end{array}$$
$$\begin{array}{r} 27 \\ \times\ 9 \\ \hline \end{array}$$
$$\begin{array}{r} 45 \\ \times\ 7 \\ \hline \end{array}$$
$$\begin{array}{r} 34 \\ \times\ 8 \\ \hline \end{array}$$

E.
$$\begin{array}{r} 95 \\ \times\ 5 \\ \hline \end{array}$$
$$\begin{array}{r} 87 \\ \times\ 3 \\ \hline \end{array}$$
$$\begin{array}{r} 47 \\ \times\ 5 \\ \hline \end{array}$$
$$\begin{array}{r} 16 \\ \times\ 8 \\ \hline \end{array}$$
$$\begin{array}{r} 77 \\ \times\ 6 \\ \hline \end{array}$$

F.
$$\begin{array}{r} 36 \\ \times\ 8 \\ \hline \end{array}$$
$$\begin{array}{r} 89 \\ \times\ 3 \\ \hline \end{array}$$
$$\begin{array}{r} 58 \\ \times\ 6 \\ \hline \end{array}$$
$$\begin{array}{r} 34 \\ \times\ 9 \\ \hline \end{array}$$
$$\begin{array}{r} 78 \\ \times\ 4 \\ \hline \end{array}$$

FS-32070 Third Grade Math Review

Name _____

Skating Through Multiplication

Multiply.

A.
$$
\begin{array}{r}
{}^{3}\ 116 \\
\times\ 6 \\
\hline
696
\end{array}
\qquad
\begin{array}{r}
161 \\
\times\ 5 \\
\hline
\end{array}
\qquad
\begin{array}{r}
125 \\
\times\ 3 \\
\hline
\end{array}
\qquad
\begin{array}{r}
238 \\
\times\ 2 \\
\hline
\end{array}
\qquad
\begin{array}{r}
243 \\
\times\ 3 \\
\hline
\end{array}
$$

B.
$$
\begin{array}{r}
484 \\
\times\ 2 \\
\hline
\end{array}
\qquad
\begin{array}{r}
107 \\
\times\ 9 \\
\hline
\end{array}
\qquad
\begin{array}{r}
171 \\
\times\ 5 \\
\hline
\end{array}
\qquad
\begin{array}{r}
216 \\
\times\ 4 \\
\hline
\end{array}
\qquad
\begin{array}{r}
217 \\
\times\ 4 \\
\hline
\end{array}
$$

C.
$$
\begin{array}{r}
106 \\
\times\ 5 \\
\hline
\end{array}
\qquad
\begin{array}{r}
246 \\
\times\ 2 \\
\hline
\end{array}
\qquad
\begin{array}{r}
219 \\
\times\ 4 \\
\hline
\end{array}
\qquad
\begin{array}{r}
119 \\
\times\ 5 \\
\hline
\end{array}
\qquad
\begin{array}{r}
283 \\
\times\ 3 \\
\hline
\end{array}
$$

D.
$$
\begin{array}{r}
108 \\
\times\ 5 \\
\hline
\end{array}
\qquad
\begin{array}{r}
231 \\
\times\ 4 \\
\hline
\end{array}
\qquad
\begin{array}{r}
429 \\
\times\ 2 \\
\hline
\end{array}
\qquad
\begin{array}{r}
407 \\
\times\ 2 \\
\hline
\end{array}
\qquad
\begin{array}{r}
272 \\
\times\ 3 \\
\hline
\end{array}
$$

E.
$$
\begin{array}{r}
427 \\
\times\ 2 \\
\hline
\end{array}
\qquad
\begin{array}{r}
109 \\
\times\ 6 \\
\hline
\end{array}
\qquad
\begin{array}{r}
218 \\
\times\ 4 \\
\hline
\end{array}
\qquad
\begin{array}{r}
329 \\
\times\ 3 \\
\hline
\end{array}
\qquad
\begin{array}{r}
124 \\
\times\ 3 \\
\hline
\end{array}
$$

F.
$$
\begin{array}{r}
724 \\
\times\ 3 \\
\hline
\end{array}
\qquad
\begin{array}{r}
205 \\
\times\ 4 \\
\hline
\end{array}
\qquad
\begin{array}{r}
352 \\
\times\ 4 \\
\hline
\end{array}
\qquad
\begin{array}{r}
627 \\
\times\ 3 \\
\hline
\end{array}
\qquad
\begin{array}{r}
283 \\
\times\ 4 \\
\hline
\end{array}
$$

FS-32070 Third Grade Math Review

Name _____

Multiplication Wizard

Multiply.

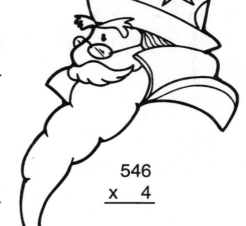

A.
```
  1
  351
x   2
  702
```

```
  372
x   3
```

```
  124
x   3
```

B.
```
  103
x   9
```

```
  249
x   4
```

```
  377
x   2
```

```
  546
x   4
```

```
  208
x   4
```

C.
```
  382
x   3
```

```
  417
x   2
```

```
  126
x   4
```

```
  238
x   3
```

```
  158
x   5
```

D.
```
  206
x   3
```

```
  324
x   3
```

```
  168
x   4
```

```
  295
x   3
```

```
  128
x   3
```

E.
```
  135
x   4
```

```
  175
x   5
```

```
  309
x   2
```

```
  218
x   4
```

```
  408
x   2
```

F.
```
  196
x   4
```

```
  319
x   2
```

```
  256
x   3
```

```
  165
x   4
```

```
  127
x   6
```

FS-32070 Third Grade Math Review

Multiplying Money

Multiply.

A.
$ 0.37
x   5
$1.85

$ 0.46
x   2

$ 1.37
x   4

$ 3.15
x   7

$ 2.05
x   4

B.
$ 0.76
x   9

$ 5.26
x   2

$ 3.14
x   3

$ 0.54
x   9

$ 2.21
x   4

C.
$ 0.15
x   8

$ 1.32
x   5

$ 0.38
x   4

$ 1.15
x   3

$ 3.72
x   4

D.
$ 0.48
x   6

$ 6.05
x   5

$ 2.20
x   4

$ 0.49
x   4

$ 0.33
x   4

E.
$ 0.54
x   3

$ 1.34
x   5

$ 0.27
x   3

$ 4.80
x   2

$ 3.54
x   4

F.
$ 7.28
x   3

$ 0.88
x   9

$ 3.20
x   6

$ 1.62
x   6

$ 5.32
x   2

FS-32070 Third Grade Math Review

A Pot of Gold

Multiply.

A.
$$\overset{5\ 4}{\$0.77} \times 7 = \$5.39$$

$2.13
x   4

$1.65
x   2

B.
$1.15
x   5

$1.05
x   6

$4.83
x   2

$0.89
x   5

$1.37
x   4

C.
$2.63
x   5

$7.14
x   2

$0.87
x   6

$0.92
x   3

$0.53
x   7

D.
$4.42
x   5

$8.07
x   3

$1.42
x   7

$0.36
x   5

$0.27
x   9

E.
$2.25
x   2

$6.85
x   3

$3.05
x   3

$1.94
x   2

$2.80
x   4

F.
$5.17
x   3

$0.85
x   9

$0.48
x   7

$0.17
x   6

$0.35
x   5

A Carousel Ride

Divide.

A.
$$5\overline{)36}$$
$$\begin{array}{r} 7R1 \\ 5\overline{)36} \\ -35 \\ \hline 1 \end{array}$$
$$5\overline{)48}$$
$$3\overline{)17}$$

B.
$$7\overline{)66}$$
$$3\overline{)25}$$
$$4\overline{)37}$$
$$6\overline{)19}$$
$$5\overline{)44}$$

C.
$$6\overline{)39}$$
$$2\overline{)17}$$
$$8\overline{)20}$$
$$4\overline{)29}$$
$$9\overline{)11}$$

D.
$$9\overline{)75}$$
$$5\overline{)33}$$
$$4\overline{)17}$$
$$7\overline{)10}$$
$$2\overline{)13}$$

E.
$$3\overline{)29}$$
$$5\overline{)27}$$
$$2\overline{)11}$$
$$3\overline{)23}$$
$$4\overline{)35}$$

FS-32070 Third Grade Math Review

Name_____
Dividing two-digit numbers
by one-digit numbers

Let's Divide

Divide.

A.
$$5)\overline{34} \quad \frac{6R4}{-30}$$
$$\frac{4}{}$$
2)$\overline{15}$　　　3)$\overline{28}$　　　5)$\overline{43}$　　　8)$\overline{44}$

B.
6)$\overline{14}$　　　7)$\overline{30}$　　　9)$\overline{15}$　　　4)$\overline{29}$　　　7)$\overline{23}$

C.
5)$\overline{37}$　　　4)$\overline{33}$　　　3)$\overline{19}$　　　6)$\overline{25}$　　　9)$\overline{20}$

D.
4)$\overline{22}$　　　3)$\overline{13}$　　　9)$\overline{10}$　　　5)$\overline{17}$　　　7)$\overline{46}$

E.
2)$\overline{11}$　　　3)$\overline{20}$　　　4)$\overline{30}$

FS-32070 Third Grade Math Review

Name _____

Step by Step

Divide.

A.   3)‾43‾   14R1
     -3
     ‾13‾
     -12
     ‾1‾

4)‾50‾   5)‾63‾

B.   2)‾35‾   3)‾82‾   5)‾72‾   4)‾67‾   2)‾39‾

C.   5)‾68‾   4)‾91‾   2)‾87‾   5)‾84‾   2)‾85‾

D.   5)‾59‾   2)‾97‾   3)‾64‾   4)‾94‾   7)‾85‾

FS-32070 Third Grade Math Review

Name_____

A River Ride

Divide.

A. $\overset{16R3}{4\overline{)67}}$ $2\overline{)51}$ $3\overline{)40}$ $5\overline{)69}$ $7\overline{)89}$

$\underline{-4}$
27
$\underline{-24}$
3

B. $5\overline{)63}$ $4\overline{)46}$ $3\overline{)43}$ $2\overline{)65}$ $6\overline{)87}$

C. $7\overline{)97}$ $4\overline{)93}$ $5\overline{)58}$ $3\overline{)71}$ $2\overline{)89}$

D. $2\overline{)73}$ $3\overline{)82}$ $4\overline{)63}$

FS-32070 Third Grade Math Review

Up, Up, and Away

Divide. If the quotient has no remainder, color the section on the hot-air balloon with the matching letter.

A.
$$
\begin{array}{r}
42 \\
3\overline{)126} \\
-12 \\
\hline
6 \\
-6 \\
\hline
0
\end{array}
$$

B. $5\overline{)456}$

C. $3\overline{)136}$

D. $4\overline{)352}$

E. $2\overline{)158}$

F. $5\overline{)217}$

G. $3\overline{)257}$

H. $4\overline{)264}$

I. $2\overline{)113}$

J. $4\overline{)114}$

K. $5\overline{)355}$

L. $6\overline{)156}$

Name_____
Dividing three-digit numbers
by one-digit numbers

Diamonds Are Forever

Divide.

 A
$$\begin{array}{r} 28 \\ 4\overline{)112} \\ -8 \\ \hline 32 \\ -32 \\ \hline 0 \end{array}$$

 B $5\overline{)105}$

 D $3\overline{)237}$

 E $4\overline{)127}$

 F $5\overline{)150}$

 I $4\overline{)195}$

 L $3\overline{)129}$

 N $5\overline{)108}$

 O $5\overline{)206}$

 S $4\overline{)179}$

 T $3\overline{)123}$

 U $4\overline{)167}$

Fill in the correct letter over each answer.
Where can you find the largest diamond in the world?

___	___	A	___	A	___	___	___	A	___	___
41R1	21R3	28	21	28	44R3	31R3	21	28	43	43

___	___	___	___	___
30	48R3	31R3	43	79

102

Off-road Division

Divide.

A.
$$\frac{198 R1}{4 \overline{)793}}$$
$$-4$$
$$39$$
$$-36$$
$$33$$
$$-32$$
$$1$$

$5 \overline{)642}$

B. $4 \overline{)865}$ $3 \overline{)756}$ $2 \overline{)678}$ $6 \overline{)856}$

C. $6 \overline{)940}$ $8 \overline{)986}$ $5 \overline{)813}$ $3 \overline{)791}$

D. $3 \overline{)971}$ $6 \overline{)789}$ $8 \overline{)899}$ $7 \overline{)926}$

FS-32070 Third Grade Math Review

Name _____
Dividing three-digit numbers
by one-digit numbers

Brushing Up on Division

Divide.

A.
$$\begin{array}{r} 123R3 \\ 7\overline{)864} \\ -7 \\ \hline 16 \\ -14 \\ \hline 24 \\ -21 \\ \hline 3 \end{array}$$

$4\overline{)567}$ $6\overline{)940}$ $2\overline{)557}$

B. $5\overline{)687}$ $3\overline{)398}$ $3\overline{)759}$ $8\overline{)889}$

C. $8\overline{)984}$ $3\overline{)891}$ $2\overline{)927}$ $6\overline{)768}$

D. $6\overline{)837}$ $4\overline{)647}$

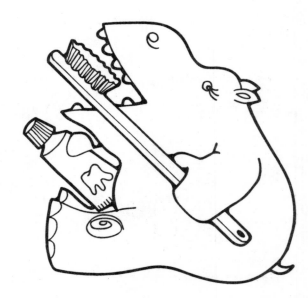

Answer Key

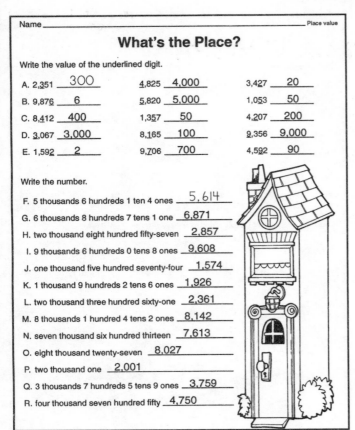

Page 1

Name _____ Place value

What's the Place?

Write the value of the underlined digit.

A. 2,351 __300__
B. 9,876 __6__
C. 8,412 __400__
D. 3,067 __3,000__
E. 1,592 __2__

4,825 __4,000__
5,820 __5,000__
1,357 __50__
8,165 __100__
9,706 __700__

3,427 __20__
1,053 __50__
4,207 __200__
9,356 __9,000__
4,592 __90__

Write the number.

F. 5 thousands 6 hundreds 1 ten 4 ones __5,614__
G. 6 thousands 8 hundreds 7 tens 1 one __6,871__
H. two thousand eight hundred fifty-seven __2,857__
I. 9 thousands 6 hundreds 0 tens 8 ones __9,608__
J. one thousand five hundred seventy-four __1,574__
K. 1 thousand 9 hundreds 2 tens 6 ones __1,926__
L. two thousand three hundred sixty-one __2,361__
M. 8 thousands 1 hundred 4 tens 2 ones __8,142__
N. seven thousand six hundred thirteen __7,613__
O. eight thousand twenty-seven __8,027__
P. two thousand one __2,001__
Q. 3 thousands 7 hundreds 5 tens 9 ones __3,759__
R. four thousand seven hundred fifty __4,750__

Page 1

Page 2

Name _____ Place value

A Builder's Dream

| 1 | 2 | 3 | 4 | 5 | 6 | 7 | 8 | 9 |

Use the numbers above to write a four-digit number on each line.

A. the greatest number __9,876__
B. the greatest number with 7 in the thousands place __7,986__
C. the least number __1,234__
D. the greatest odd number __9,875__
E. the least odd number __1,235__
F. the greatest number with all even digits __8,642__
G. the least number with all odd digits __1,357__

Write the number or numbers that have the given value.

3,584 8,426 9,075 7,531 2,001

H. greater than 4,000 __7,531; 8,426; 9,075__
I. less than 2,500 __2,001__
J. between 7,000 and 8,999 __7,531 ; 8,426__
K. less than 5,000 __2,001 ; 3,584__
L. between 1,000 and 2,999 __2,001__
M. greater than 9,000 __9,075__

Page 2

Page 3

Name _____ Place value

Puzzle Place

Complete the puzzle by writing the correct number for each clue.

Across

A. six hundred fourteen thousand two hundred ninety-seven
E. three hundred twenty-two
F. five hundred twenty-nine
G. four thousand eighteen
H. twenty
J. ten
K. twenty-seven
L. forty-seven
M. nineteen thousand six hundred thirty-four
P. seven thousand eight hundred twenty
R. seven thousand four hundred thirty-five
S. sixty-two thousand five hundred thirty-one

Down

A. six hundred twenty-one
B. one hundred twenty-eight
C. nine hundred fifty-two
D. seven hundred twenty
E. thirty thousand ninety-four
G. four thousand one hundred seventeen
I. eight thousand seven hundred one
K. twenty-four
L. four hundred twenty-three
N. sixty-three
O. thirty-five
P. seventy-two
Q. eighty-five

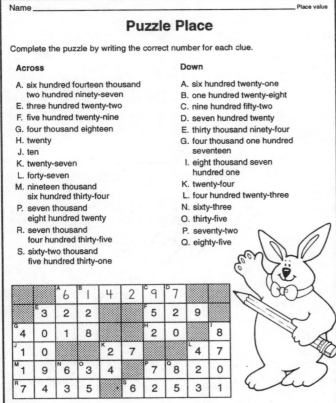

Page 3

Page 4

Name _____ Place value

Soaring With Numbers

Write the value of the underlined digit.

A. 468,930 __60,000__
B. 16,497 __10,000__
C. 54,082 __4,000__
D. 231,856 __200,000__
E. 842,031 __2,000__
F. 602,937 __600,000__
G. 869,235 __800,000__
H. 237,981 __30,000__
I. 49,625 __9,000__
J. 986,417 __900,000__

84,261 __200__
375,829 __300,000__
493,650 __90,000__
587,465 __7,000__
98,714 __90,000__
850,674 __600__
926,504 __20,000__
679,815 __10__
498,024 __400,000__
523,127 __20,000__

Write the numbers from the kites where the digit 6 has the given value.

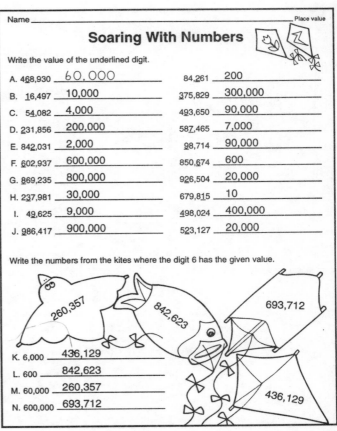

260,357   842,623   693,712   436,129

K. 6,000 __436,129__
L. 600 __842,623__
M. 60,000 __260,357__
N. 600,000 __693,712__

Page 4

105

FS-32070 Third Grade Math Review

Answer Key

Rounding to the nearest ten and hundred

Round About

Round to the nearest ten.

A. 29	30	48	50	82	80
B. 17	20	31	30	63	60
C. 92	90	59	60	23	20
D. 88	90	12	10	94	90
E. 26	30	37	40	72	70
F. 56	60	66	70	74	70
G. 45	50	88	90	39	40

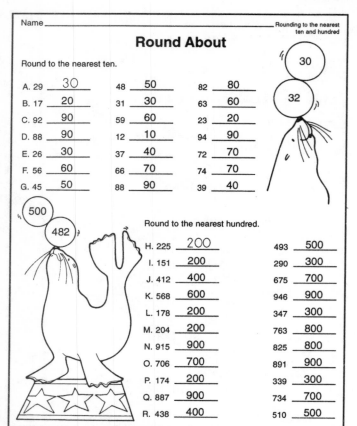

Round to the nearest hundred.

H. 225	200	493	500
I. 151	200	290	300
J. 412	400	675	700
K. 568	600	946	900
L. 178	200	347	300
M. 204	200	763	800
N. 915	900	825	800
O. 706	700	891	900
P. 174	200	339	300
Q. 887	900	734	700
R. 438	400	510	500

Page 5

Rounding to the nearest ten and hundred

Rounding Numbers

Circle the numbers that would round to the first number in each row.

A. 80	(76)	(83)	74	91	(75)	85
B. 40	53	47	(37)	(44)	(35)	(39)
C. 60	(63)	(58)	54	(64)	(59)	65
D. 50	55	(45)	(49)	(53)	(48)	57
E. 30	(32)	(26)	41	(29)	(34)	35
F. 90	(91)	96	(86)	(94)	(85)	79

Circle the numbers that would round to the last number in each row.

G.	(257)	352	403	(295)	(335)	300
H.	(519)	645	619	581	(467)	500
I.	(875)	(914)	(850)	950	(923)	900
J.	743	865	(768)	(802)	(848)	800
K.	193	(137)	98	(143)	(129)	100

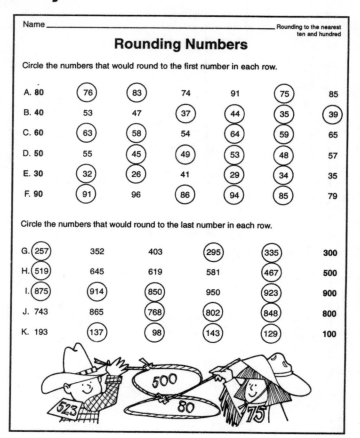

Page 6

Addition facts to 18

Catch the Facts

Add.

A.	5 +7 = 12	7 +9 = 16	8 +4 = 12	6 +9 = 15	9 +4 = 13	8 +9 = 17
B.	2 +8 = 10	9 +5 = 14	3 +8 = 11	6 +6 = 12	5 +5 = 10	2 +9 = 11
C.	9 +8 = 17	6 +4 = 10	4 +7 = 11	9 +7 = 16	3 +9 = 12	6 +8 = 14
D.	5 +9 = 14	4 +8 = 12	5 +8 = 13	9 +9 = 18	3 +6 = 9	8 +5 = 13

Write the missing number.

E. 7 + [8] = 15   4 + [5] = 9   9 + [3] = 12   6 + [7] = 13

F. 4 + [6] = 10   6 + [5] = 11   8 + [6] = 14   4 + [4] = 8

G. 7 + [6] = 13   3 + [7] = 10   7 + [7] = 14   8 + [8] = 16

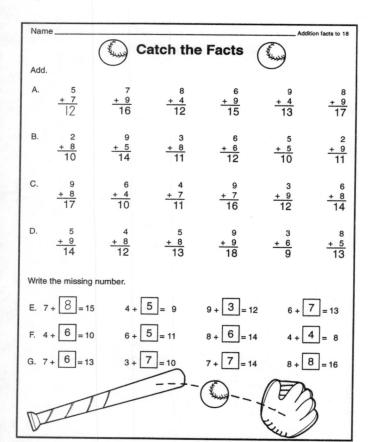

Page 7

Addition facts to 18

Beat the Clock

Add.

A.	5 +6 = 11	8 +7 = 15	3 +4 = 7	9 +6 = 15	7 +9 = 16	8 +5 = 13
B.	6 +8 = 14	9 +4 = 13	5 +9 = 14	8 +9 = 17	4 +7 = 11	3 +9 = 12
C.	7 +3 = 10	7 +8 = 15	2 +7 = 9	5 +7 = 12	8 +8 = 16	7 +5 = 12
D.	6 +6 = 12	4 +4 = 8	5 +8 = 13	7 +4 = 11	2 +9 = 11	8 +6 = 14

Write the missing number.

E. 9 + [5] = 14   6 + [9] = 15   6 + [7] = 13

F. 3 + [7] = 10   4 + [6] = 10   9 + [7] = 16

G. 6 + [4] = 10   3 + [8] = 11   7 + [6] = 13

H. 8 + [4] = 12   9 + [8] = 17   6 + [5] = 11

I. 9 + [9] = 18   7 + [7] = 14   4 + [9] = 13

Page 8

Answer Key

Over the Rainbow

Name _____ Two-digit addition with regrouping once

Add.

A. 29 +13 = 42	45 +16 = 61	85 + 9 = 94	53 +18 = 71	42 +29 = 71
B. 45 +39 = 84	57 +18 = 75	69 +12 = 81	48 +38 = 86	26 +26 = 52
C. 37 +18 = 55	29 +56 = 85	43 +37 = 80	71 +19 = 90	47 +49 = 96
D. 18 +18 = 36	52 +18 = 70	66 + 7 = 73	49 +15 = 64	28 +63 = 91
E. 48 +17 = 65	46 +26 = 72	55 +25 = 80	73 +17 = 90	37 +15 = 52

F. 72 +18 = 90 | 24 +66 = 90 | 63 +17 = 80

G. 25 +27 = 52 | 43 +19 = 62 | 39 +28 = 67

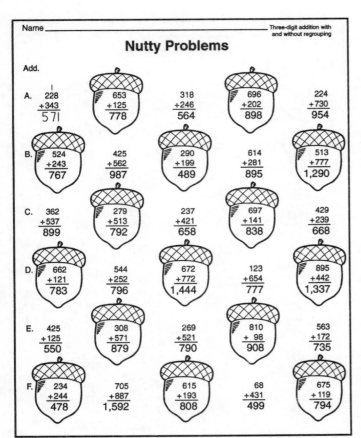

Page 9

A Quilt of Problems

Name _____ Two-digit addition with regrouping once

Add.

A. 38 +16 = 54	47 +13 = 60	29 +29 = 58	35 +25 = 60	13 +49 = 62
B. 19 +13 = 32	17 +18 = 35	19 +26 = 45	28 +27 = 55	37 +35 = 72
C. 46 +46 = 92	53 +29 = 82	27 +45 = 72	59 + 3 = 62	28 +24 = 52
D. 27 +23 = 50	37 +18 = 55	48 +12 = 60	59 + 6 = 65	39 +31 = 70
E. 47 +33 = 80	39 +39 = 78	59 +17 = 76	49 +25 = 74	35 +57 = 92
F. 68 +22 = 90	18 +15 = 33	24 +29 = 53	59 +14 = 73	65 +28 = 93

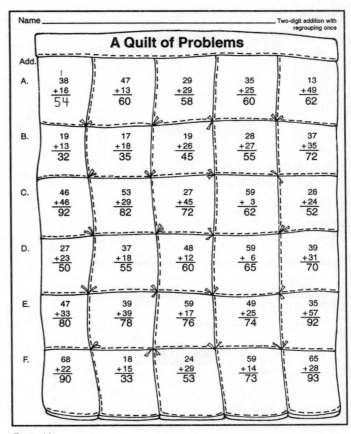

Page 10

Nutty Problems

Name _____ Three-digit addition with and without regrouping

Add.

A. 228 +343 = 571	653 +125 = 778	318 +246 = 564	696 +202 = 898	224 +730 = 954
B. 524 +243 = 767	425 +562 = 987	290 +199 = 489	614 +281 = 895	513 +777 = 1,290
C. 362 +537 = 899	279 +513 = 792	237 +421 = 658	697 +141 = 838	429 +239 = 668
D. 662 +121 = 783	544 +252 = 796	672 +772 = 1,444	123 +654 = 777	895 +442 = 1,337
E. 425 +125 = 550	308 +571 = 879	269 +521 = 790	810 + 98 = 908	563 +172 = 735
F. 234 +244 = 478	705 +887 = 1,592	615 +193 = 808	68 +431 = 499	675 +119 = 794

Page 11

Going Bananas Over Addition

Name _____ Three-digit addition with and without regrouping

Add.

A. 233 +534 = 767	763 +152 = 915	617 +119 = 736	135 +531 = 666	90 +879 = 969
B. 345 +190 = 535	500 +123 = 623	793 +142 = 935	547 +129 = 676	416 + 83 = 499
C. 816 +147 = 963	135 +243 = 378	353 +122 = 475	227 +106 = 333	809 + 90 = 899
D. 435 +534 = 969	193 +342 = 535	806 + 82 = 888	377 +171 = 548	843 + 25 = 868
E. 264 +462 = 726	313 +342 = 655	631 +136 = 767	794 + 94 = 888	642 +319 = 961
F. 125 +101 = 226	365 +432 = 797	272 +416 = 688		
G. 444 +190 = 634	463 +365 = 828	797 +202 = 999		

Page 12

FS-32070 Third Grade Math Review

Answer Key

Name _____

Sum Path

Add.

A. 128 +113 **241**	B. 431 +129 560	C. 170 + 66 236	D. 367 +143 510	E. 183 + 52 235
F. 494 +265 759	G. 327 +392 719	H. 368 +464 832	I. 379 +214 593	J. 842 + 63 905
K. 766 +135 901	L. 543 +288 831	M. 450 +150 600	N. 372 +349 721	O. 175 +386 561
P. 674 +175 849	Q. 346 +195 541	R. 389 +102 491	S. 478 +135 613	T. 283 +149 432

Start at the toy store. Draw a line to connect the sums from problems A to T.

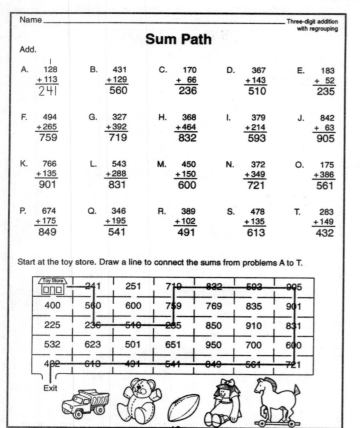

Page 13

Name _____

All Together Now

Add.

I 235 +196 **431**	U 336 +365 701	F 264 +573 837	G 536 +156 692	D 836 + 47 883
W 236 +580 816	E 152 +356 508	O 676 +251 927	N 237 +628 865	X 299 +350 649
J 484 +292 776	U 321 +496 817	K 538 +191 729	R 214 +576 790	G 372 +244 616
R 903 + 48 951	B 793 +198 991	H 487 +305 792	L 679 +246 925	O 587 +406 993

Fill in the correct letter over each answer.
What is everything in the world doing at the same time?

G R O W I N G   O L D E R
692 790 927 816 431 865 616   993 925 883 508 951

Page 14

Name _____

Great Bear Sums

Add. Circle the greatest sum in each row.

A. 4,567 +1,360 **5,927**	4,835 +4,124 8,959	3,973 +1,024 4,997	8,243 + 961 **(9,204)**
B. 5,832 +1,745 7,577	3,467 +2,981 6,448	3,009 +1,996 5,005	6,253 +1,626 **(7,879)**
C. 6,175 +2,834 **(9,009)**	4,862 +2,105 6,967	5,032 +1,304 6,336	3,125 +5,862 8,987
D. 5,152 +4,734 **(9,886)**	6,371 + 983 7,354	2,484 +3,105 5,589	1,396 +4,609 6,005
E. 2,166 +4,730 6,896	2,156 + 239 2,395	8,653 +1,047 9,700	9,137 + 842 **(9,979)**
F. 2,786 +2,017 4,803	8,197 + 102 **(8,299)**		
G. 6,854 +1,023 7,877	8,266 + 424 **(8,690)**		

Add

Page 15

Name _____

Prickly Problems

Add.

A. 4,357 +3,932 **8,289**	2,352 +6,435 8,787	5,075 + 123 5,198	7,256 +1,982 9,238
B. 4,353 +2,405 6,758	5,705 +1,836 7,541	7,216 +1,642 8,858	3,852 +3,975 7,827
C. 5,036 +2,823 7,859	6,852 +1,607 8,459	2,834 +5,165 7,999	972 +7,846 8,818
D. 4,358 +3,483 7,841	5,215 +2,732 7,947	2,852 +6,056 8,908	1,207 +3,402 4,609
E. 5,697 +1,903 7,600	7,285 + 614 7,899	6,343 +1,756 8,099	5,216 +1,893 7,109
F. 4,367 +2,332 6,699	842 +7,198 8,040		
G. 6,097 + 903 7,000	5,318 +2,460 7,778		

Page 16

Answer Key

Name _____   Column addition

Triple Hitter

Add.

A.	45 37 +14 **96**	28 30 +19 **77**	70 62 +57 **189**	81 29 +43 **153**	53 15 +22 **90**
B.	75 63 +50 **188**	75 28 +19 **122**	54 95 +77 **226**	75 19 +68 **162**	46 97 +13 **156**
C.	12 89 +35 **136**	54 17 +47 **118**	45 32 +23 **100**	62 78 +97 **237**	77 86 +18 **181**
D.	528 306 +217 **1,051**	293 687 +319 **1,299**	846 797 +904 **2,547**	496 185 +206 **887**	360 875 +496 **1,731**
E.	876 257 +18 **1,151**	865 79 +491 **1,435**	746 829 +688 **2,263**		
F.	826 835 +206 **1,867**	496 87 +142 **725**	221 107 +322 **650**		

Name _____   Column addition

Shooting Stars

Add.

A.	75 28 +20 **123**	29 38 +27 **94**	88 39 +47 **174**	58 49 +26 **133**	23 97 +29 **149**
B.	24 66 +31 **121**	69 12 +14 **95**		32 87 +43 **162**	19 86 +43 **148**
C.		27 50 +38 **115**	36 45 +89 **170**	97 35 +22 **154**	27 18 +67 **112**
D.	235 196 +267 **698**	327 450 +118 **895**	836 245 +589 **1,670**		297 635 +342 **1,274**
E.	687 618 +167 **1,472**		830 103 +428 **1,361**	490 374 +221 **1,085**	865 231 +814 **1,910**
F.	625 103 +134 **862**	504 190 +847 **1,541**	680 208 +112 **1,000**	124 179 +813 **1,116**	

Name _____   Addition with regrouping

Totally Cool

Add.

A.	347 +2,693 **3,040**	562 +975 **1,537**	3,652 +79 **3,731**	4,385 +1,823 **6,208**
B.	2,682 +1,937 **4,619**	972 +83 **1,055**	6,096 +997 **7,093**	3,862 +98 **3,960**
C.	4,975 +127 **5,102**	367 +2,896 **3,263**	4,065 +3,985 **8,050**	250 +475 **725**
D.	375 +987 **1,362**	2,856 +85 **2,941**	9,349 +467 **9,816**	1,126 +7,956 **9,082**
E.	2,223 +85 **2,308**	537 +187 **724**	8,516 +1,095 **9,611**	2,987 +563 **3,550**
F.	8,675 +296 **8,971**	5,093 +1,923 **7,016**	6,781 +969 **7,750**	378 +699 **1,077**

Name _____   Addition with regrouping

"Sum"mer Fun

Add.

A.	4,865 +1,987 **6,852**	132 +989 **1,121**	4,096 +875 **4,971**	328 +597 **925**
B.	3,117 +983 **4,100**	5,870 +562 **6,432**	1,835 +6,926 **8,761**	8,458 +97 **8,555**
C.	4,497 +2,536 **7,033**	1,432 +976 **2,408**	8,158 +87 **8,245**	858 +596 **1,454**
D.	386 +509 **895**	2,983 +5,087 **8,070**	5,681 +299 **5,980**	943 +876 **1,819**
E.	6,888 +219 **7,107**	4,098 +923 **5,021**	6,948 +2,065 **9,013**	3,342 +2,964 **6,306**
F.			5,106 +2,983 **8,089**	486 +926 **1,412**
G.			1,836 +890 **2,726**	8,817 +827 **9,644**

     FS-32070 Third Grade Math Review

Answer Key

In the Ballpark

Round to the nearest ten. Estimate the sum.

A.
32 +68 → 30 + 70 = 100
79 +67 → 80 + 70 = 150
56 +39 → 60 + 40 = 100

B.
81 +58 → 80 + 60 = 140
17 +46 → 20 + 50 = 70
73 +66 → 70 + 70 = 140

C.
78 +43 → 80 + 40 = 120
62 +91 → 60 + 90 = 150
43 +48 → 40 + 50 = 90

Round to the nearest hundred. Estimate the sum.

D.
342 +287 → 300 + 300 = 600
625 +303 → 600 + 300 = 900
297 +502 → 300 + 500 = 800

E.
569 +119 → 600 + 100 = 700
749 +475 → 700 + 500 = 1,200
685 +804 → 700 + 800 = 1,500

F.
483 +915 → 500 + 900 = 1,400
858 +743 → 900 + 700 = 1,600

Page 21

Rounding Up and Down

Round to the nearest ten. Estimate the sum.

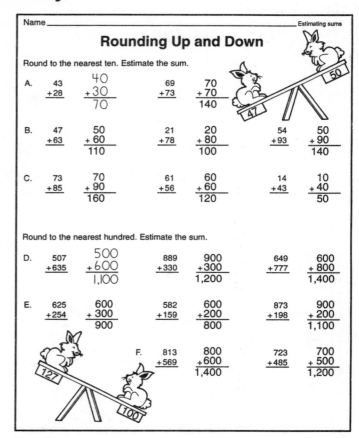

A.
43 +28 → 40 + 30 = 70
69 +73 → 70 + 70 = 140

B.
47 +63 → 50 + 60 = 110
21 +78 → 20 + 80 = 100
54 +93 → 50 + 90 = 140

C.
73 +85 → 70 + 90 = 160
61 +56 → 60 + 60 = 120
14 +43 → 10 + 40 = 50

Round to the nearest hundred. Estimate the sum.

D.
507 +635 → 500 + 600 = 1,100
889 +330 → 900 + 300 = 1,200
649 +777 → 600 + 800 = 1,400

E.
625 +254 → 600 + 300 = 900
582 +159 → 600 + 200 = 800
873 +198 → 900 + 200 = 1,100

F.
813 +569 → 800 + 600 = 1,400
723 +485 → 700 + 500 = 1,200

Page 22

Dancing Differences

Subtract.

A.
12 − 8 = 4
15 − 3 = 12
11 − 3 = 8
10 − 7 = 3
8 − 4 = 4
9 − 6 = 3

B.
16 − 9 = 7
13 − 7 = 6
15 − 6 = 9
17 − 8 = 9
14 − 7 = 7
13 − 4 = 9

C.
16 − 8 = 8
14 − 5 = 9
13 − 6 = 7
16 − 7 = 9
11 − 8 = 3
10 − 5 = 5

D.
18 − 9 = 9
15 − 7 = 8
14 − 6 = 8
17 − 9 = 8
10 − 4 = 6
9 − 4 = 5

Write the missing number.

E. 15 − [8] = 7   10 − [8] = 2   12 − [6] = 6

F. 11 − [5] = 6   13 − [8] = 5   14 − [6] = 8

G. 10 − [3] = 7   12 − [5] = 7   12 − [3] = 9

H. 14 − [5] = 9   11 − [6] = 5   12 − [4] = 8

I. 12 − [7] = 5   13 − [5] = 8   13 − [4] = 9

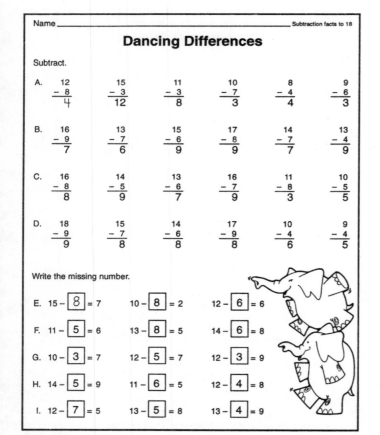

Page 23

Subtraction Action

Subtract.

A.
13 − 9 = 4
10 − 3 = 7
16 − 8 = 8
10 − 6 = 4
11 − 6 = 5
12 − 6 = 6

B.
15 − 6 = 9
11 − 9 = 2
16 − 9 = 7
10 − 4 = 6
12 − 3 = 9
14 − 5 = 9

C.
18 − 9 = 9
17 − 9 = 8
11 − 8 = 3
12 − 4 = 8
13 − 8 = 5
12 − 7 = 5

D.
17 − 8 = 9
16 − 7 = 9
10 − 5 = 5
11 − 4 = 7
12 − 5 = 7
13 − 4 = 9

Write the missing number.

E. 10 − [2] = 8   15 − [7] = 8   11 − [7] = 4   12 − [9] = 3

F. 11 − [5] = 6   13 − [7] = 6   13 − [5] = 8   11 − [3] = 8

G. 12 − [8] = 4   14 − [7] = 7   14 − [8] = 6   10 − [8] = 2

Page 24

Answer Key

Clowning With Subtraction

Two-digit subtraction with and without regrouping

Subtract.

A.	3 10	63	80	66	52
	4̶0̶ −12 = 28	−29 = 34	−27 = 53	−37 = 29	−14 = 38

B.	93 −77 = 16	95 −56 = 39	63 −11 = 52	83 − 9 = 74	59 −30 = 29

C.	68 −52 = 16	71 − 5 = 66	43 −25 = 18	43 −18 = 25	55 −36 = 19

D.	81 −12 = 69	94 −76 = 18	85 −27 = 58	96 −38 = 58	83 −22 = 61

E.	80 −67 = 13	46 −43 = 3	75 −27 = 48	78 −65 = 13	82 −76 = 6

F.	59 −42 = 17	32 −18 = 14	56 −48 = 8		

Balloons: 75, 78, 36

G.	79 −34 = 45	91 −65 = 26	65 −32 = 33

Page 25

Cuddly Subtraction

Two-digit subtraction with and without regrouping

Subtract.

A.	4 11	62	83	71	76
	5̶1̶ −18 = 33	−13 = 49	− 4 = 79	−47 = 24	−55 = 21

B.	97 −27 = 70	50 −18 = 32	76 −29 = 47	88 −36 = 52	66 −57 = 9

C.	72 −45 = 27	95 −22 = 73	33 −18 = 15	59 −32 = 27	62 −53 = 9

D.	88 −32 = 56	79 −20 = 59	60 −19 = 41	67 − 9 = 58	88 −79 = 9

E.	50 −38 = 12	66 −54 = 12	83 −36 = 47	43 −22 = 21	80 −44 = 36

F.	91 −27 = 64	85 −43 = 42	75 −65 = 10	42 − 8 = 34	76 −43 = 33

Page 26

Carnival Math

Three-digit subtraction with regrouping once

Subtract.

A.	6 11	243	926		
	6̶7̶1̶ −134 = 537	− 82 = 161	−432 = 494		

B.	764 −138 = 626	134 − 74 = 60	825 −195 = 630		

C.	328 −295 = 33	692 − 38 = 654	892 −153 = 739	725 − 92 = 633	683 −456 = 227

D.	425 −180 = 245	627 −392 = 235	821 −340 = 481	866 −293 = 573	892 −375 = 517

E.	835 −362 = 473	952 −825 = 127	378 −293 = 85	610 −301 = 309	842 −290 = 552

F.	738 −193 = 545	625 −491 = 134	816 −385 = 431	293 −147 = 146	514 −283 = 231

G.	773 −444 = 329	628 −575 = 53	275 − 67 = 208	970 −865 = 105	766 −392 = 374

Page 27

Submerge Into Subtraction

Three-digit subtraction with regrouping once

Subtract.

A.	6 13	244	817	753	671
	7̶3̶2̶ −141 = 591	− 73 = 171	−382 = 435	−272 = 481	−333 = 338

B.	819 −584 = 235	135 − 84 = 51	781 −127 = 654	980 −252 = 728	734 −571 = 163

C.	438 −229 = 209	644 −526 = 118	249 −168 = 81	384 − 92 = 292	848 −663 = 185

D.	776 −285 = 491	641 −350 = 291	338 −253 = 85	162 − 48 = 114	871 −258 = 613

E.	982 −374 = 608	625 −431 = 194	474 − 81 = 393	880 −452 = 428	467 −284 = 183

F.	463 −181 = 282	861 −425 = 436	767 −249 = 518	627 −275 = 352	935 −341 = 594

Page 28

FS-32070 Third Grade Math Review

Answer Key

Page 29

Name _____

Three-digit subtraction with regrouping twice

Pick of the Litter

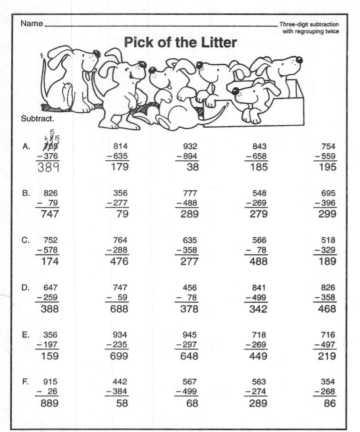

Subtract.

A.
765 − 376 = 389
814 − 635 = 179
932 − 894 = 38
843 − 658 = 185
754 − 559 = 195

B.
826 − 79 = 747
356 − 277 = 79
777 − 488 = 289
548 − 269 = 279
695 − 396 = 299

C.
752 − 578 = 174
764 − 288 = 476
635 − 358 = 277
566 − 78 = 488
518 − 329 = 189

D.
647 − 259 = 388
747 − 59 = 688
456 − 78 = 378
841 − 499 = 342
826 − 358 = 468

E.
356 − 197 = 159
934 − 235 = 699
945 − 297 = 648
718 − 269 = 449
716 − 497 = 219

F.
915 − 26 = 889
442 − 384 = 58
567 − 499 = 68
563 − 274 = 289
354 − 268 = 86

Page 29

Page 30

Name _____

Three-digit subtraction with regrouping twice

Playful Problems

Subtract.

A.
563 − 178 = 385
862 − 495 = 367
713 − 198 = 515
326 − 129 = 197
553 − 68 = 485

B.
723 − 427 = 296
261 − 87 = 174
546 − 379 = 167
743 − 295 = 448
983 − 597 = 386

C.
764 − 466 = 298
793 − 195 = 598
541 − 166 = 375
552 − 278 = 274
824 − 295 = 529

D.
353 − 268 = 85
777 − 378 = 399
964 − 175 = 789
915 − 496 = 419
821 − 722 = 99

E.
121 − 99 = 22
828 − 749 = 79
875 − 387 = 488
871 − 296 = 575
321 − 233 = 88

F.
727 − 438 = 289
874 − 396 = 478
532 − 166 = 366
516 − 329 = 187
974 − 596 = 378

Page 30

Page 31

Name _____

Three-digit subtraction with zeros

Crystal Ball

Subtract. Fill in the correct letter over each answer.
What is it that you cannot see, but is always ahead of you?

T H E   F U T U R E
308 58 391   405 384 245 6 355 67

T	S	E	J	H
809 − 564 = 245	180 − 42 = 138	705 − 314 = 391	430 − 214 = 216	903 − 845 = 58

B	G	U	C	D
510 − 76 = 434	805 − 688 = 117	530 − 146 = 384	603 − 478 = 125	800 − 496 = 304

U	K	L	T	E
304 − 298 = 6	400 − 291 = 109	900 − 567 = 333	604 − 296 = 308	700 − 633 = 67

P	M	F	N	R
906 − 609 = 297	800 − 743 = 57	603 − 198 = 405	408 − 285 = 123	702 − 347 = 355

Page 31

Page 32

Name _____

Three-digit subtraction with zeros

A Secret Word

Subtract. Color the boxes with differences of 500 or less.

A.				
504 − 275 = 229	905 − 386 = 519	700 − 267 = 433	806 − 259 = 547	703 − 255 = 448
B. 508 − 369 = 139	908 − 79 = 829	805 − 473 = 332	701 − 48 = 653	800 − 246 = 554
C. 307 − 268 = 39	440 − 328 = 112	805 − 485 = 320	904 − 265 = 639	506 − 175 = 331
D. 706 − 380 = 326	650 − 67 = 583	703 − 208 = 495	900 − 312 = 588	700 − 248 = 452
E. 902 − 789 = 113	800 − 284 = 516	801 − 357 = 444	603 − 27 = 576	804 − 399 = 405
F. 706 − 407 = 299	607 − 59 = 548	700 − 295 = 405	906 − 329 = 577	808 − 343 = 465

What is the secret word? __HI__

Page 32

Answer Key

Name _____

Soar Through the Air

Four-digit subtraction
with regrouping once

Subtract.

A. 4,374   6,636   5,295
  −2,256   −5,429   −2,883
  2,118   1,207   2,412

B. 5,843   8,254   3,173
  −3,561   −5,832   −1,158
  2,282   2,422   2,015

C. 3,680   5,663   6,202   3,768
  −1,342   −4,249   −4,051   −1,924
  2,338   1,414   2,151   1,844

D. 6,170   4,431   2,846   5,845
  −3,158   −2,630   −1,593   −1,629
  3,012   1,801   1,253   4,216

E. 7,059   8,868   7,908   5,291
  −5,343   −6,339   −3,150   −3,078
  1,716   2,529   4,758   2,213

F. 8,367   7,746   8,692   5,446
  −2,623   −5,385   −2,486   −1,329
  5,744   2,361   6,206   4,117

G. 7,665   9,456   5,243   3,543
  −2,754   −4,173   −1,823   −1,493
  4,911   5,283   3,420   2,050

Page 33

Name _____

Highflying Kites

Four-digit subtraction
with regrouping once

Subtract.

A. 7,886   4,374
  −4,079   −2,861
  3,807   1,513

B. 8,364   8,746
  −7,229   −2,485
  1,135   6,261

C. 6,234   6,670   9,538   9,327
  −1,520   −6,341   −9,275   −1,257
  4,714   329   263   8,070

D. 5,759   7,287   6,876   5,314
  −1,939   −1,532   −2,459   −2,504
  3,820   5,755   4,417   2,810

E. 6,805   7,620   5,832   6,473
  −5,472   −3,280   −5,160   −4,392
  1,333   4,340   672   2,081

F. 6,274   1,723   4,139   7,623
  −3,463   −1,504   −2,618   −5,408
  2,811   219   1,521   2,215

G. 6,924   5,296   9,423   7,258
  −3,452   −4,167   −5,418   −4,735
  3,472   1,129   4,005   2,523

Page 34

Name _____

Mountain Climbing

Four-digit subtraction with
regrouping more than once

Subtract.

A. 6,253   1,537   3,731   6,208
  −1,896   − 875   − 879   −1,823
  4,357   622   2,852   4,385

B. 4,055   4,619   6,955   7,003
  −1,967   −2,682   −3,983   − 907
  2,088   1,937   2,972   6,096

C. 3,960   9,325   5,000   6,913
  − 98   −8,750   −3,975   −3,578
  3,862   575   1,025   3,335

D. 8,050   9,725   1,362   9,806
  −3,985   − 942   − 987   − 457
  4,065   8,783   375   9,349

E. 6,208   3,001   8,012   8,771
  −3,985   −1,503   −2,751   − 96
  2,223   1,498   5,261   8,675

F. 4,723   6,075
  −1,952   −2,738
  2,771   3,337

G. 7,018   9,007
  −1,923   −8,669
  5,095   338

Page 35

Name _____

Investigating

Four-digit subtraction with
regrouping more than once

Subtract.

A. 3,305   6,140   3,623   5,704
  −1,391   −4,350   −1,496   − 932
  1,914   1,790   2,127   4,772

B. 6,200   1,405   8,240   8,047
  − 172   − 67   −5,639   −6,239
  6,028   1,338   2,601   1,808

C. 3,200   7,042   9,000   5,403
  −1,195   −5,453   −4,170   −2,794
  2,005   1,589   4,830   2,609

D. 6,088   4,896   5,000   8,075
  − 295   −1,497   −1,818   − 79
  5,793   3,399   3,182   7,996

E. 4,309   7,058   1,003   9,637
  −1,428   −5,466   − 757   − 829
  2,881   1,592   246   8,808

F. 9,009   1,600   8,083   6,022
  − 567   −1,189   −2,989   −5,926
  8,442   411   5,094   96

Page 36

113

Answer Key

Best Estimate

Name _____ Estimating differences

Round to the nearest ten. Estimate the difference.

A.
38 40
−12 −10
    30

57 60
−28 −30
    30

81 80
−49 −50
    30

B.
93 90
−27 −30
    60

41 40
−19 −20
    20

85 90
−46 −50
    40

C.
53 50
−31 −30
    20

42 40
−28 −30
    10

92 90
−68 −70
    20

Round to the nearest hundred. Estimate the difference.

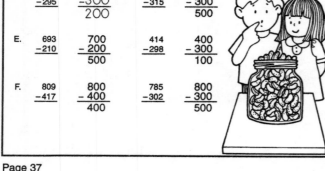

200?   370?

D.
508 500
−295 −300
     200

789 800
−315 −300
     500

E.
693 700
−210 −200
     500

414 400
−298 −300
     100

F.
809 800
−417 −400
     400

785 800
−302 −300
     500

Page 37

A Pile of Gifts

Name _____ Estimating differences

Round to the nearest ten. Estimate the difference.

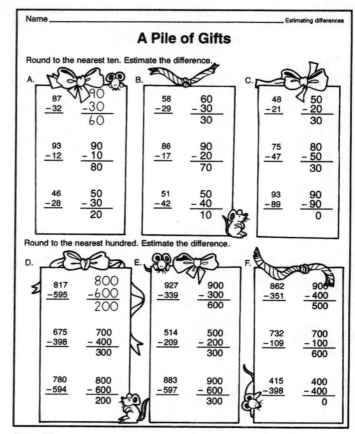

A.
87 90
−32 −30
    60

93 90
−12 −10
    80

46 50
−28 −30
    20

B.
58 60
−29 −30
    30

86 90
−17 −20
    70

51 50
−42 −40
    10

C.
48 50
−21 −20
    30

75 80
−47 −50
    30

93 90
−89 −90
     0

Round to the nearest hundred. Estimate the difference.

D.
817 800
−595 −600
     200

675 700
−398 −400
     300

780 800
−594 −600
     200

E.
927 900
−339 −300
     600

514 500
−209 −200
     300

883 900
−597 −600
     300

F.
862 900
−351 −400
     500

732 700
−109 −100
     600

415 400
−398 −400
     0

Page 38

Count by Groups

Name _____ Multiplying by 2 and 5

Write a number sentence for each picture.

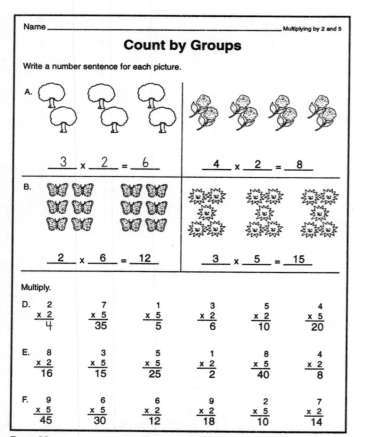

A.   __3__ x __2__ = __6__   __4__ x __2__ = __8__

B.   __2__ x __6__ = __12__   __3__ x __5__ = __15__

Multiply.

D.
2   7   1   3   5   4
x 2 x 5 x 5 x 2 x 2 x 5
 4   35   5   6   10   20

E.
8   3   5   1   8   4
x 2 x 5 x 5 x 2 x 5 x 2
16   15   25   2   40   8

F.
9   6   6   9   2   7
x 5 x 5 x 2 x 2 x 5 x 2
45   30   12   18   10   14

Page 39

Multiplication Snakes

Name _____ Multiplying by 2 and 5

Multiply.

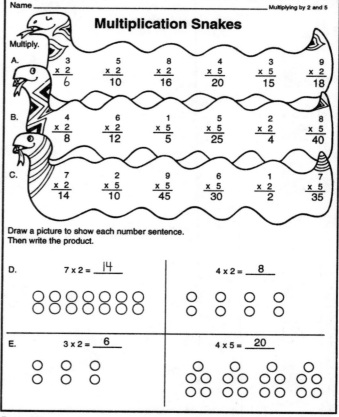

A.
3   5   8   4   3   9
x 2 x 2 x 2 x 5 x 5 x 2
 6   10   16   20   15   18

B.
4   6   1   5   2   8
x 2 x 2 x 5 x 5 x 2 x 5
 8   12   5   25   4   40

C.
7   2   9   6   1   7
x 2 x 5 x 5 x 5 x 2 x 5
14   10   45   30   2   35

Draw a picture to show each number sentence.
Then write the product.

D.   7 x 2 = __14__   4 x 2 = __8__

E.   3 x 2 = __6__   4 x 5 = __20__

Page 40

FS-32070 Third Grade Math Review

Answer Key

Name _____

Multiplication March

Multiply.

A.
$\begin{array}{r} 5 \\ \times 3 \\ \hline 15 \end{array}$
$\begin{array}{r} 4 \\ \times 4 \\ \hline 16 \end{array}$
$\begin{array}{r} 7 \\ \times 3 \\ \hline 21 \end{array}$
$\begin{array}{r} 4 \\ \times 3 \\ \hline 12 \end{array}$
$\begin{array}{r} 2 \\ \times 4 \\ \hline 8 \end{array}$
$\begin{array}{r} 3 \\ \times 1 \\ \hline 3 \end{array}$

B.
$\begin{array}{r} 6 \\ \times 3 \\ \hline 18 \end{array}$
$\begin{array}{r} 1 \\ \times 3 \\ \hline 3 \end{array}$
$\begin{array}{r} 3 \\ \times 4 \\ \hline 12 \end{array}$
$\begin{array}{r} 8 \\ \times 4 \\ \hline 32 \end{array}$
$\begin{array}{r} 9 \\ \times 3 \\ \hline 27 \end{array}$
$\begin{array}{r} 7 \\ \times 4 \\ \hline 28 \end{array}$

C.
$\begin{array}{r} 9 \\ \times 4 \\ \hline 36 \end{array}$
$\begin{array}{r} 2 \\ \times 3 \\ \hline 6 \end{array}$
$\begin{array}{r} 4 \\ \times 6 \\ \hline 24 \end{array}$
$\begin{array}{r} 5 \\ \times 4 \\ \hline 20 \end{array}$
$\begin{array}{r} 3 \\ \times 5 \\ \hline 15 \end{array}$
$\begin{array}{r} 4 \\ \times 2 \\ \hline 8 \end{array}$

D.
$\begin{array}{r} 8 \\ \times 3 \\ \hline 24 \end{array}$
$\begin{array}{r} 1 \\ \times 4 \\ \hline 4 \end{array}$
$\begin{array}{r} 3 \\ \times 3 \\ \hline 9 \end{array}$
$\begin{array}{r} 6 \\ \times 4 \\ \hline 24 \end{array}$
$\begin{array}{r} 4 \\ \times 5 \\ \hline 20 \end{array}$
$\begin{array}{r} 3 \\ \times 2 \\ \hline 6 \end{array}$

E. $3 \times 9 = 27$ $4 \times 3 = 12$ $4 \times 7 = 28$

F. $4 \times 9 = 36$ $3 \times 8 = 24$ $3 \times 7 = 21$

G. $3 \times 6 = 18$ $4 \times 8 = 32$ $4 \times 6 = 24$

Page 41

Name _____

Packaged Products

Multiply.

A. $4 \times 3 = 12$ $5 \times 4 = 20$ $6 \times 4 = 24$

B. $8 \times 3 = 24$ $1 \times 4 = 4$ $1 \times 3 = 3$

C. $9 \times 4 = 36$ $5 \times 3 = 15$ $7 \times 3 = 21$

D. $3 \times 3 = 9$ $8 \times 4 = 32$ $3 \times 4 = 12$

E. $2 \times 3 = 6$ $6 \times 3 = 18$ $7 \times 4 = 28$

F. $4 \times 4 = 16$ $2 \times 4 = 8$ $9 \times 3 = 27$

G.
$\begin{array}{r} 4 \\ \times 7 \\ \hline 28 \end{array}$
$\begin{array}{r} 9 \\ \times 3 \\ \hline 27 \end{array}$
$\begin{array}{r} 4 \\ \times 4 \\ \hline 16 \end{array}$
$\begin{array}{r} 3 \\ \times 5 \\ \hline 15 \end{array}$
$\begin{array}{r} 2 \\ \times 3 \\ \hline 6 \end{array}$
$\begin{array}{r} 3 \\ \times 7 \\ \hline 21 \end{array}$

H.
$\begin{array}{r} 3 \\ \times 8 \\ \hline 24 \end{array}$
$\begin{array}{r} 1 \\ \times 4 \\ \hline 4 \end{array}$
$\begin{array}{r} 4 \\ \times 3 \\ \hline 12 \end{array}$
$\begin{array}{r} 7 \\ \times 3 \\ \hline 21 \end{array}$
$\begin{array}{r} 4 \\ \times 8 \\ \hline 32 \end{array}$
$\begin{array}{r} 3 \\ \times 3 \\ \hline 9 \end{array}$

I.
$\begin{array}{r} 2 \\ \times 4 \\ \hline 8 \end{array}$
$\begin{array}{r} 3 \\ \times 6 \\ \hline 18 \end{array}$
$\begin{array}{r} 4 \\ \times 9 \\ \hline 36 \end{array}$
$\begin{array}{r} 8 \\ \times 4 \\ \hline 32 \end{array}$
$\begin{array}{r} 4 \\ \times 5 \\ \hline 20 \end{array}$
$\begin{array}{r} 3 \\ \times 2 \\ \hline 6 \end{array}$

J.
$\begin{array}{r} 3 \\ \times 9 \\ \hline 27 \end{array}$
$\begin{array}{r} 4 \\ \times 2 \\ \hline 8 \end{array}$
$\begin{array}{r} 4 \\ \times 6 \\ \hline 24 \end{array}$
$\begin{array}{r} 3 \\ \times 1 \\ \hline 3 \end{array}$

Page 42

Name _____

Hop to It!

Multiply.

A. $4 \times 1 = 4$ $6 \times 0 = 0$ $1 \times 9 = 9$

B. $7 \times 1 = 7$ $4 \times 0 = 0$ $6 \times 1 = 6$

C. $9 \times 0 = 0$ $1 \times 1 = 1$ $1 \times 7 = 7$

D. $9 \times 1 = 9$ $8 \times 0 = 0$ $3 \times 1 = 3$

E. $0 \times 7 = 0$ $2 \times 1 = 2$ $8 \times 1 = 8$

F. $2 \times 0 = 0$ $5 \times 1 = 5$ $0 \times 0 = 0$

G.
$\begin{array}{r} 0 \\ \times 4 \\ \hline 0 \end{array}$
$\begin{array}{r} 1 \\ \times 8 \\ \hline 8 \end{array}$
$\begin{array}{r} 1 \\ \times 2 \\ \hline 2 \end{array}$
$\begin{array}{r} 0 \\ \times 5 \\ \hline 0 \end{array}$
$\begin{array}{r} 1 \\ \times 0 \\ \hline 0 \end{array}$
$\begin{array}{r} 0 \\ \times 8 \\ \hline 0 \end{array}$

H.
$\begin{array}{r} 1 \\ \times 3 \\ \hline 3 \end{array}$
$\begin{array}{r} 3 \\ \times 0 \\ \hline 0 \end{array}$
$\begin{array}{r} 1 \\ \times 4 \\ \hline 4 \end{array}$
$\begin{array}{r} 0 \\ \times 6 \\ \hline 0 \end{array}$
$\begin{array}{r} 1 \\ \times 9 \\ \hline 9 \end{array}$
$\begin{array}{r} 5 \\ \times 0 \\ \hline 0 \end{array}$

I.
$\begin{array}{r} 0 \\ \times 9 \\ \hline 0 \end{array}$
$\begin{array}{r} 1 \\ \times 6 \\ \hline 6 \end{array}$
$\begin{array}{r} 0 \\ \times 3 \\ \hline 0 \end{array}$
$\begin{array}{r} 7 \\ \times 0 \\ \hline 0 \end{array}$
$\begin{array}{r} 0 \\ \times 2 \\ \hline 0 \end{array}$
$\begin{array}{r} 1 \\ \times 5 \\ \hline 5 \end{array}$

Page 43

Name _____

Merry-Go-Round Multiplication

Multiply.

A.
$\begin{array}{r} 3 \\ \times 0 \\ \hline 0 \end{array}$
$\begin{array}{r} 4 \\ \times 1 \\ \hline 4 \end{array}$
$\begin{array}{r} 5 \\ \times 0 \\ \hline 0 \end{array}$
$\begin{array}{r} 8 \\ \times 0 \\ \hline 0 \end{array}$
$\begin{array}{r} 9 \\ \times 1 \\ \hline 9 \end{array}$
$\begin{array}{r} 2 \\ \times 1 \\ \hline 2 \end{array}$

B.
$\begin{array}{r} 6 \\ \times 0 \\ \hline 0 \end{array}$
$\begin{array}{r} 3 \\ \times 1 \\ \hline 3 \end{array}$
$\begin{array}{r} 9 \\ \times 0 \\ \hline 0 \end{array}$
$\begin{array}{r} 7 \\ \times 1 \\ \hline 7 \end{array}$
$\begin{array}{r} 8 \\ \times 1 \\ \hline 8 \end{array}$
$\begin{array}{r} 1 \\ \times 0 \\ \hline 0 \end{array}$

C.
$\begin{array}{r} 1 \\ \times 1 \\ \hline 1 \end{array}$
$\begin{array}{r} 5 \\ \times 1 \\ \hline 5 \end{array}$
$\begin{array}{r} 7 \\ \times 0 \\ \hline 0 \end{array}$
$\begin{array}{r} 4 \\ \times 0 \\ \hline 0 \end{array}$
$\begin{array}{r} 2 \\ \times 0 \\ \hline 0 \end{array}$
$\begin{array}{r} 6 \\ \times 1 \\ \hline 6 \end{array}$

D. $0 \times 1 = 0$ $0 \times 6 = 0$ $1 \times 2 = 2$

E. $1 \times 7 = 7$ $1 \times 5 = 5$ $0 \times 2 = 0$

F. $0 \times 4 = 0$ $0 \times 5 = 0$ $1 \times 3 = 3$

G. $1 \times 6 = 6$ $1 \times 4 = 4$ $0 \times 0 = 0$

H. $1 \times 8 = 8$ $0 \times 3 = 0$ $1 \times 9 = 9$

I. $0 \times 7 = 0$ $0 \times 8 = 0$ $0 \times 9 = 0$

Page 44

FS-32070 Third Grade Math Review

Answer Key

Name _____

Crack the Code

Multiply. Fill in the correct letter over each answer.
Why is the letter B hot?

B E C A U S E I T
14 49 18 35 54 56 49 28 42

M A K E S O I L B O I L
63 35 36 49 56 21 28 48 14 21 28 48

B $\begin{array}{r} 2 \\ \times 7 \\ \hline 14 \end{array}$	C $\begin{array}{r} 3 \\ \times 6 \\ \hline 18 \end{array}$	F $\begin{array}{r} 6 \\ \times 5 \\ \hline 30 \end{array}$	G $\begin{array}{r} 1 \\ \times 7 \\ \hline 7 \end{array}$	A $\begin{array}{r} 5 \\ \times 7 \\ \hline 35 \end{array}$	D $\begin{array}{r} 4 \\ \times 6 \\ \hline 24 \end{array}$
F $\begin{array}{r} 5 \\ \times 6 \\ \hline 30 \end{array}$	H $\begin{array}{r} 0 \\ \times 7 \\ \hline 0 \end{array}$	I $\begin{array}{r} 4 \\ \times 7 \\ \hline 28 \end{array}$	J $\begin{array}{r} 1 \\ \times 6 \\ \hline 6 \end{array}$	M $\begin{array}{r} 9 \\ \times 7 \\ \hline 63 \end{array}$	O $\begin{array}{r} 3 \\ \times 7 \\ \hline 21 \end{array}$
P $\begin{array}{r} 2 \\ \times 6 \\ \hline 12 \end{array}$	K $\begin{array}{r} 6 \\ \times 6 \\ \hline 36 \end{array}$	S $\begin{array}{r} 8 \\ \times 7 \\ \hline 56 \end{array}$	T $\begin{array}{r} 7 \\ \times 6 \\ \hline 42 \end{array}$	H $\begin{array}{r} 0 \\ \times 6 \\ \hline 0 \end{array}$	A $\begin{array}{r} 7 \\ \times 5 \\ \hline 35 \end{array}$
U $\begin{array}{r} 9 \\ \times 6 \\ \hline 54 \end{array}$	T $\begin{array}{r} 6 \\ \times 7 \\ \hline 42 \end{array}$	O $\begin{array}{r} 7 \\ \times 3 \\ \hline 21 \end{array}$	E $\begin{array}{r} 7 \\ \times 7 \\ \hline 49 \end{array}$	D $\begin{array}{r} 6 \\ \times 4 \\ \hline 24 \end{array}$	L $\begin{array}{r} 8 \\ \times 6 \\ \hline 48 \end{array}$

Page 45

Name _____

Play Ball!

Multiply.

A. $\begin{array}{r} 5 \\ \times 6 \\ \hline 30 \end{array}$ $\begin{array}{r} 4 \\ \times 7 \\ \hline 28 \end{array}$ $\begin{array}{r} 6 \\ \times 7 \\ \hline 42 \end{array}$ $\begin{array}{r} 2 \\ \times 6 \\ \hline 12 \end{array}$ $\begin{array}{r} 1 \\ \times 7 \\ \hline 7 \end{array}$ $\begin{array}{r} 7 \\ \times 6 \\ \hline 42 \end{array}$

B. $\begin{array}{r} 4 \\ \times 6 \\ \hline 24 \end{array}$ $\begin{array}{r} 3 \\ \times 7 \\ \hline 21 \end{array}$ $\begin{array}{r} 6 \\ \times 6 \\ \hline 36 \end{array}$ $\begin{array}{r} 1 \\ \times 6 \\ \hline 6 \end{array}$ $\begin{array}{r} 5 \\ \times 7 \\ \hline 35 \end{array}$ $\begin{array}{r} 9 \\ \times 6 \\ \hline 54 \end{array}$

C. $\begin{array}{r} 2 \\ \times 7 \\ \hline 14 \end{array}$ $\begin{array}{r} 3 \\ \times 6 \\ \hline 18 \end{array}$ $\begin{array}{r} 8 \\ \times 6 \\ \hline 48 \end{array}$ $\begin{array}{r} 7 \\ \times 7 \\ \hline 49 \end{array}$ $\begin{array}{r} 8 \\ \times 7 \\ \hline 56 \end{array}$ $\begin{array}{r} 9 \\ \times 7 \\ \hline 63 \end{array}$

Match each fact with its product.

D. 6 x 4 — 28 → 24
E. 7 x 4 — 12 → 28
F. 7 x 3 — 24 → 21
G. 6 x 2 — 21 → 12
H. 6 x 5 — 63 → 30
I. 7 x 9 — 54 → 63
J. 6 x 9 — 30 → 54

7 x 5 — 48 → 35
6 x 8 — 18 → 48
7 x 8 — 35 → 56
6 x 3 — 14 → 18
7 x 2 — 56 → 14
6 x 7 — 6 → 42
1 x 6 — 42 → 6

Page 46

Name _____

Diving Into Multiplication

Multiply.

A. $\begin{array}{r} 4 \\ \times 8 \\ \hline 32 \end{array}$ $\begin{array}{r} 2 \\ \times 9 \\ \hline 18 \end{array}$ $\begin{array}{r} 4 \\ \times 9 \\ \hline 36 \end{array}$ $\begin{array}{r} 7 \\ \times 9 \\ \hline 63 \end{array}$ $\begin{array}{r} 6 \\ \times 8 \\ \hline 48 \end{array}$ $\begin{array}{r} 8 \\ \times 4 \\ \hline 32 \end{array}$

B. $\begin{array}{r} 8 \\ \times 9 \\ \hline 72 \end{array}$ $\begin{array}{r} 3 \\ \times 8 \\ \hline 24 \end{array}$ $\begin{array}{r} 8 \\ \times 9 \\ \hline 72 \end{array}$ $\begin{array}{r} 8 \\ \times 8 \\ \hline 64 \end{array}$ $\begin{array}{r} 5 \\ \times 9 \\ \hline 45 \end{array}$ $\begin{array}{r} 8 \\ \times 5 \\ \hline 40 \end{array}$

C. $\begin{array}{r} 3 \\ \times 9 \\ \hline 27 \end{array}$ $\begin{array}{r} 9 \\ \times 8 \\ \hline 72 \end{array}$ $\begin{array}{r} 7 \\ \times 8 \\ \hline 56 \end{array}$ $\begin{array}{r} 5 \\ \times 8 \\ \hline 40 \end{array}$ $\begin{array}{r} 6 \\ \times 9 \\ \hline 54 \end{array}$ $\begin{array}{r} 9 \\ \times 9 \\ \hline 81 \end{array}$

Match each fact with its product.

D. 8 x 3 — 9
E. 9 x 5 — 24
F. 9 x 1 — 48
G. 8 x 6 — 18
H. 9 x 7 — 45
I. 9 x 2 — 56
J. 8 x 7 — 63

Page 47

Name _____

Multiplication Bumper Cars

Multiply.

A. $\begin{array}{r} 2 \\ \times 8 \\ \hline 16 \end{array}$ $\begin{array}{r} 9 \\ \times 9 \\ \hline 81 \end{array}$ $\begin{array}{r} 7 \\ \times 8 \\ \hline 56 \end{array}$ $\begin{array}{r} 8 \\ \times 9 \\ \hline 72 \end{array}$ $\begin{array}{r} 5 \\ \times 8 \\ \hline 40 \end{array}$ $\begin{array}{r} 8 \\ \times 6 \\ \hline 48 \end{array}$

B. $\begin{array}{r} 9 \\ \times 8 \\ \hline 72 \end{array}$ $\begin{array}{r} 4 \\ \times 8 \\ \hline 32 \end{array}$ $\begin{array}{r} 6 \\ \times 9 \\ \hline 54 \end{array}$ $\begin{array}{r} 8 \\ \times 4 \\ \hline 32 \end{array}$ $\begin{array}{r} 9 \\ \times 4 \\ \hline 36 \end{array}$ $\begin{array}{r} 3 \\ \times 8 \\ \hline 24 \end{array}$

C. $\begin{array}{r} 8 \\ \times 8 \\ \hline 64 \end{array}$ $\begin{array}{r} 9 \\ \times 6 \\ \hline 54 \end{array}$ $\begin{array}{r} 4 \\ \times 9 \\ \hline 36 \end{array}$ $\begin{array}{r} 8 \\ \times 3 \\ \hline 24 \end{array}$ $\begin{array}{r} 9 \\ \times 7 \\ \hline 63 \end{array}$ $\begin{array}{r} 6 \\ \times 8 \\ \hline 48 \end{array}$

D. $\begin{array}{r} 9 \\ \times 2 \\ \hline 18 \end{array}$ $\begin{array}{r} 9 \\ \times 5 \\ \hline 45 \end{array}$ $\begin{array}{r} 8 \\ \times 9 \\ \hline 72 \end{array}$ $\begin{array}{r} 1 \\ \times 8 \\ \hline 8 \end{array}$ $\begin{array}{r} 9 \\ \times 9 \\ \hline 81 \end{array}$ $\begin{array}{r} 8 \\ \times 5 \\ \hline 40 \end{array}$

E. $\begin{array}{r} 8 \\ \times 7 \\ \hline 56 \end{array}$ $\begin{array}{r} 8 \\ \times 1 \\ \hline 8 \end{array}$ $\begin{array}{r} 2 \\ \times 9 \\ \hline 18 \end{array}$ $\begin{array}{r} 0 \\ \times 9 \\ \hline 0 \end{array}$ $\begin{array}{r} 7 \\ \times 8 \\ \hline 56 \end{array}$ $\begin{array}{r} 9 \\ \times 1 \\ \hline 9 \end{array}$

F. $\begin{array}{r} 5 \\ \times 9 \\ \hline 45 \end{array}$ $\begin{array}{r} 8 \\ \times 2 \\ \hline 16 \end{array}$ $\begin{array}{r} 3 \\ \times 9 \\ \hline 27 \end{array}$ $\begin{array}{r} 7 \\ \times 9 \\ \hline 63 \end{array}$ $\begin{array}{r} 8 \\ \times 5 \\ \hline 40 \end{array}$ $\begin{array}{r} 9 \\ \times 3 \\ \hline 27 \end{array}$

Page 48

FS-32070 Third Grade Math Review

Answer Key

Facts and Figures

Complete the table.

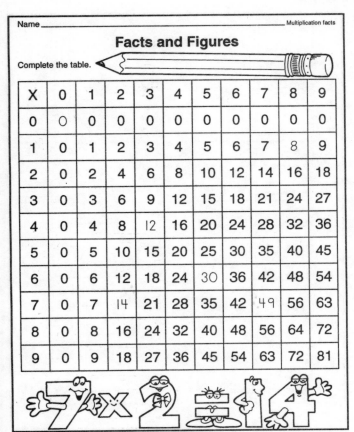

X	0	1	2	3	4	5	6	7	8	9
0	0	0	0	0	0	0	0	0	0	0
1	0	1	2	3	4	5	6	7	8	9
2	0	2	4	6	8	10	12	14	16	18
3	0	3	6	9	12	15	18	21	24	27
4	0	4	8	12	16	20	24	28	32	36
5	0	5	10	15	20	25	30	35	40	45
6	0	6	12	18	24	30	36	42	48	54
7	0	7	14	21	28	35	42	49	56	63
8	0	8	16	24	32	40	48	56	64	72
9	0	9	18	27	36	45	54	63	72	81

Page 49

Input–Output

Complete the tables.

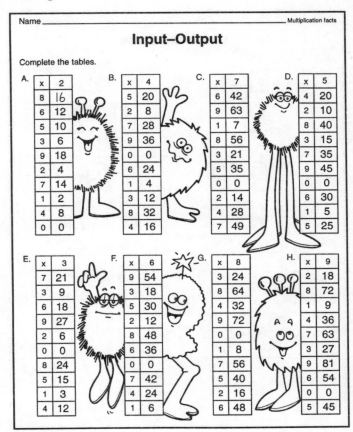

A.
x	2
8	16
6	12
5	10
3	6
9	18
2	4
7	14
1	2
4	8
0	0

B.
x	4
5	20
2	8
7	28
9	36
0	0
6	24
1	4
3	12
8	32
4	16

C.
x	7
6	42
9	63
1	7
8	56
3	21
5	35
0	0
2	14
4	28
7	49

D.
x	5
4	20
2	10
8	40
3	15
7	35
9	45
0	0
6	30
1	5
5	25

E.
x	3
7	21
3	9
6	18
9	27
2	6
0	0
8	24
5	15
1	3
4	12

F.
x	6
9	54
3	18
5	30
2	12
8	48
6	36
0	0
7	42
4	24
1	6

G.
x	8
3	24
8	64
4	32
9	72
0	0
1	8
7	56
5	40
2	16
6	48

H.
x	9
2	18
8	72
1	9
4	36
7	63
3	27
9	81
6	54
0	0
5	45

Page 50

Same Size Sets

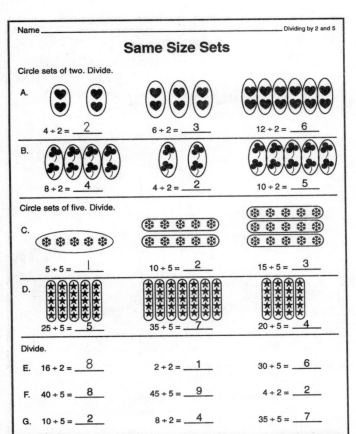

Circle sets of two. Divide.

A. 4 ÷ 2 = 2   6 ÷ 2 = 3   12 ÷ 2 = 6

B. 8 ÷ 2 = 4   4 ÷ 2 = 2   10 ÷ 2 = 5

Circle sets of five. Divide.

C. 5 ÷ 5 = 1   10 ÷ 5 = 2   15 ÷ 5 = 3

D. 25 ÷ 5 = 5   35 ÷ 5 = 7   20 ÷ 5 = 4

Divide.

E. 16 ÷ 2 = 8   2 ÷ 2 = 1   30 ÷ 5 = 6

F. 40 ÷ 5 = 8   45 ÷ 5 = 9   4 ÷ 2 = 2

G. 10 ÷ 5 = 2   8 ÷ 2 = 4   35 ÷ 5 = 7

Page 51

Winter Fun

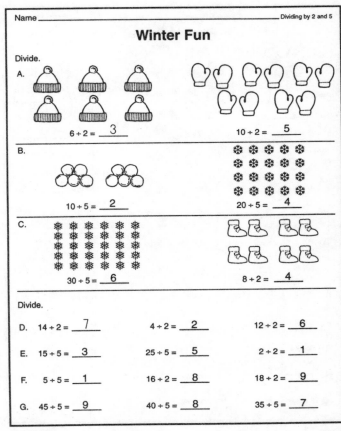

Divide.

A. 6 ÷ 2 = 3   10 ÷ 2 = 5

B. 10 ÷ 5 = 2   20 ÷ 5 = 4

C. 30 ÷ 5 = 6   8 ÷ 2 = 4

Divide.

D. 14 ÷ 2 = 7   4 ÷ 2 = 2   12 ÷ 2 = 6

E. 15 ÷ 5 = 3   25 ÷ 5 = 5   2 ÷ 2 = 1

F. 5 ÷ 5 = 1   16 ÷ 2 = 8   18 ÷ 2 = 9

G. 45 ÷ 5 = 9   40 ÷ 5 = 8   35 ÷ 5 = 7

Page 52

FS-32070 Third Grade Math Review

Answer Key

Building Blocks

Divide.

A.

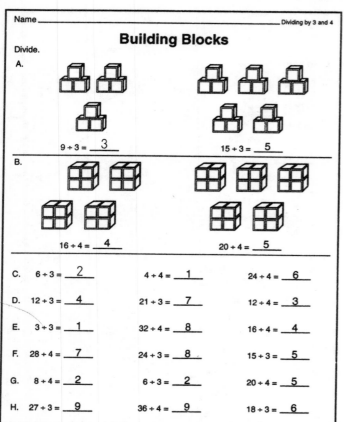

$9 \div 3 = \underline{3}$ $15 \div 3 = \underline{5}$

B.

$16 \div 4 = \underline{4}$ $20 \div 4 = \underline{5}$

C. $6 \div 3 = \underline{2}$	$4 \div 4 = \underline{1}$	$24 \div 4 = \underline{6}$
D. $12 \div 3 = \underline{4}$	$21 \div 3 = \underline{7}$	$12 \div 4 = \underline{3}$
E. $3 \div 3 = \underline{1}$	$32 \div 4 = \underline{8}$	$16 \div 4 = \underline{4}$
F. $28 \div 4 = \underline{7}$	$24 \div 3 = \underline{8}$	$15 \div 3 = \underline{5}$
G. $8 \div 4 = \underline{2}$	$6 \div 3 = \underline{2}$	$20 \div 4 = \underline{5}$
H. $27 \div 3 = \underline{9}$	$36 \div 4 = \underline{9}$	$18 \div 3 = \underline{6}$

Buttons and Bows

Divide.

A.

$6 \div 3 = \underline{2}$ $9 \div 3 = \underline{3}$ $12 \div 3 = \underline{4}$

B.

$8 \div 4 = \underline{2}$ $12 \div 4 = \underline{3}$ $4 \div 4 = \underline{1}$

C. $6 \div 3 = \underline{2}$	$15 \div 3 = \underline{5}$	$8 \div 4 = \underline{2}$
D. $12 \div 3 = \underline{4}$	$24 \div 4 = \underline{6}$	$12 \div 4 = \underline{3}$
E. $24 \div 3 = \underline{8}$	$18 \div 3 = \underline{6}$	$21 \div 3 = \underline{7}$

F. $3\overline{)9}^{\,3}$ $3\overline{)15}^{\,5}$ $4\overline{)8}^{\,2}$ $4\overline{)16}^{\,4}$ $3\overline{)21}^{\,7}$

G. $4\overline{)28}^{\,7}$ $4\overline{)20}^{\,5}$ $3\overline{)27}^{\,9}$ $4\overline{)24}^{\,6}$ $3\overline{)24}^{\,8}$

H. $3\overline{)18}^{\,6}$ $3\overline{)6}^{\,2}$ $4\overline{)32}^{\,8}$ $4\overline{)36}^{\,9}$ $3\overline{)3}^{\,1}$

Flower Power

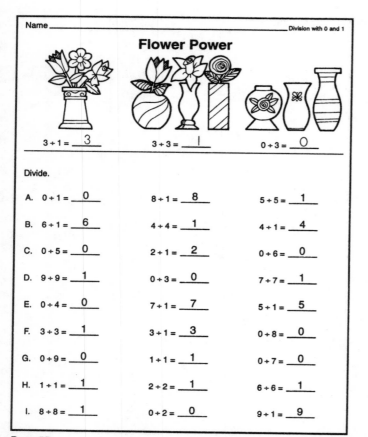

$3 \div 1 = \underline{3}$ $3 \div 3 = \underline{1}$ $0 \div 3 = \underline{0}$

Divide.

A. $0 \div 1 = \underline{0}$	$8 \div 1 = \underline{8}$	$5 \div 5 = \underline{1}$
B. $6 \div 1 = \underline{6}$	$4 \div 4 = \underline{1}$	$4 \div 1 = \underline{4}$
C. $0 \div 5 = \underline{0}$	$2 \div 1 = \underline{2}$	$0 \div 6 = \underline{0}$
D. $9 \div 9 = \underline{1}$	$0 \div 3 = \underline{0}$	$7 \div 7 = \underline{1}$
E. $0 \div 4 = \underline{0}$	$7 \div 1 = \underline{7}$	$5 \div 1 = \underline{5}$
F. $3 \div 3 = \underline{1}$	$3 \div 1 = \underline{3}$	$0 \div 8 = \underline{0}$
G. $0 \div 9 = \underline{0}$	$1 \div 1 = \underline{1}$	$0 \div 7 = \underline{0}$
H. $1 \div 1 = \underline{1}$	$2 \div 2 = \underline{1}$	$6 \div 6 = \underline{1}$
I. $8 \div 8 = \underline{1}$	$0 \div 2 = \underline{0}$	$9 \div 1 = \underline{9}$

Cookie Capers

Divide.

A. $1\overline{)9}^{\,9}$ $1\overline{)8}^{\,8}$ $1\overline{)7}^{\,7}$ $1\overline{)6}^{\,6}$ $1\overline{)5}^{\,5}$

B. $1\overline{)4}^{\,4}$ $1\overline{)3}^{\,3}$ $1\overline{)2}^{\,2}$ $1\overline{)1}^{\,1}$ $1\overline{)0}^{\,0}$

C. $9\overline{)9}^{\,1}$ $8\overline{)8}^{\,1}$ $7\overline{)7}^{\,1}$ $6\overline{)6}^{\,1}$ $5\overline{)5}^{\,1}$

D. $4\overline{)4}^{\,1}$ $3\overline{)3}^{\,1}$ $2\overline{)2}^{\,1}$ $1\overline{)1}^{\,1}$ $0\overline{)0}^{\,0}$

E. $9\overline{)0}^{\,0}$ $8\overline{)0}^{\,0}$ $7\overline{)0}^{\,0}$ $6\overline{)0}^{\,0}$

F. $5\overline{)0}^{\,0}$ $4\overline{)0}^{\,0}$ $3\overline{)0}^{\,0}$ $2\overline{)0}^{\,0}$

G. What is the quotient when a number is divided by itself?
     one

H. What is the quotient when a number is divided by 1?
     whatever that number is

I. What is the quotient when 0 is divided by a number?
     zero

FS-32070 Third Grade Math Review

Answer Key

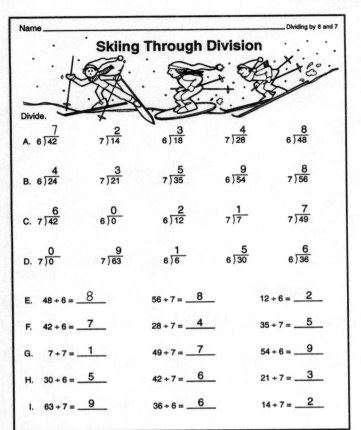

Skiing Through Division

Divide.

A. $6\overline{)42}$ = 7 $7\overline{)14}$ = 2 $6\overline{)18}$ = 3 $7\overline{)28}$ = 4 $6\overline{)48}$ = 8

B. $6\overline{)24}$ = 4 $7\overline{)21}$ = 3 $7\overline{)35}$ = 5 $6\overline{)54}$ = 9 $7\overline{)56}$ = 8

C. $7\overline{)42}$ = 6 $6\overline{)0}$ = 0 $6\overline{)12}$ = 2 $7\overline{)7}$ = 1 $7\overline{)49}$ = 7

D. $7\overline{)0}$ = 0 $7\overline{)63}$ = 9 $6\overline{)6}$ = 1 $6\overline{)30}$ = 5 $6\overline{)36}$ = 6

E. $48 \div 6 = 8$ $56 \div 7 = 8$ $12 \div 6 = 2$

F. $42 \div 6 = 7$ $28 \div 7 = 4$ $35 \div 7 = 5$

G. $7 \div 7 = 1$ $49 \div 7 = 7$ $54 \div 6 = 9$

H. $30 \div 6 = 5$ $42 \div 7 = 6$ $21 \div 7 = 3$

I. $63 \div 7 = 9$ $36 \div 6 = 6$ $14 \div 7 = 2$

Page 57

Kangaroo Power

Divide.

A. $28 \div 7 = 4$ $14 \div 7 = 2$ $6 \div 1 = 6$

B. $18 \div 6 = 3$ $42 \div 6 = 7$ $21 \div 7 = 3$

C. $0 \div 7 = 0$ $56 \div 7 = 8$ $24 \div 6 = 4$

D. $35 \div 7 = 5$ $54 \div 6 = 9$ $0 \div 6 = 0$

E. $7 \div 7 = 1$ $12 \div 6 = 2$ $30 \div 6 = 5$

F. $42 \div 7 = 6$ $36 \div 6 = 6$ $63 \div 7 = 9$

G. $48 \div 6 = 8$ $49 \div 7 = 7$ $14 \div 7 = 2$

H. $6\overline{)6}$ = 1 $6\overline{)12}$ = 2 $7\overline{)0}$ = 0 $7\overline{)42}$ = 6 $6\overline{)30}$ = 5

I. $7\overline{)21}$ = 3 $6\overline{)24}$ = 4 $7\overline{)35}$ = 5

J. $6\overline{)48}$ = 8 $7\overline{)63}$ = 9 $7\overline{)56}$ = 8

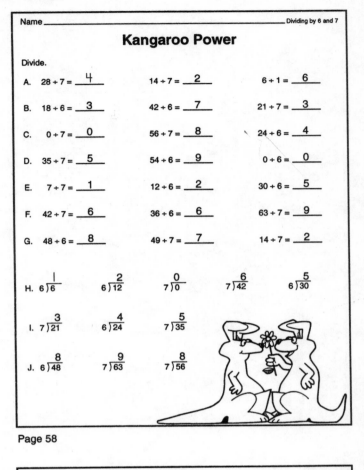

Page 58

Skydiving Fun

Divide.

A. $8 \div 8 = 1$ $40 \div 8 = 5$ $24 \div 8 = 3$

B. $16 \div 8 = 2$ $56 \div 8 = 7$ $48 \div 8 = 6$

C. $32 \div 8 = 4$ $72 \div 8 = 9$ $64 \div 8 = 8$

D. $18 \div 9 = 2$ $9 \div 9 = 1$ $72 \div 9 = 8$

E. $36 \div 9 = 4$ $27 \div 9 = 3$ $63 \div 9 = 7$

F. $54 \div 9 = 6$ $45 \div 9 = 5$ $81 \div 9 = 9$

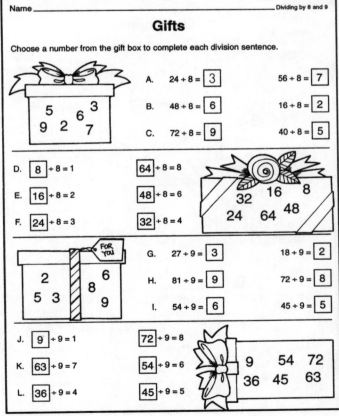

G. $9\overline{)0}$ = 0 $9\overline{)18}$ = 2 $9\overline{)27}$ = 3

H. $8\overline{)8}$ = 1 $9\overline{)81}$ = 9 $8\overline{)16}$ = 2

I. $8\overline{)32}$ = 4 $9\overline{)54}$ = 6 $9\overline{)72}$ = 8

J. $8\overline{)64}$ = 8 $8\overline{)24}$ = 3 $9\overline{)36}$ = 4 $8\overline{)56}$ = 7 $8\overline{)48}$ = 6

Page 59

Gifts

Choose a number from the gift box to complete each division sentence.

5 3
9 2 6 7

A. $24 \div 8 = \boxed{3}$ $56 \div 8 = \boxed{7}$

B. $48 \div 8 = \boxed{6}$ $16 \div 8 = \boxed{2}$

C. $72 \div 8 = \boxed{9}$ $40 \div 8 = \boxed{5}$

D. $\boxed{8} \div 8 = 1$ $\boxed{64} \div 8 = 8$

E. $\boxed{16} \div 8 = 2$ $\boxed{48} \div 8 = 6$

F. $\boxed{24} \div 8 = 3$ $\boxed{32} \div 8 = 4$

32 16 8
24 64 48

2 6
5 3 8 9

G. $27 \div 9 = \boxed{3}$ $18 \div 9 = \boxed{2}$

H. $81 \div 9 = \boxed{9}$ $72 \div 9 = \boxed{8}$

I. $54 \div 9 = \boxed{6}$ $45 \div 9 = \boxed{5}$

J. $\boxed{9} \div 9 = 1$ $\boxed{72} \div 9 = 8$

K. $\boxed{63} \div 9 = 7$ $\boxed{54} \div 9 = 6$

L. $\boxed{36} \div 9 = 4$ $\boxed{45} \div 9 = 5$

9 54 72
36 45 63

Page 60

FS-32070 Third Grade Math Review

Answer Key

Fishing for Quotients

Division facts

Divide. Color the fish.

Key			
0 or 1	red	6	green
2 or 3	orange	7	blue
4 or 5	yellow	8	purple
		9	brown

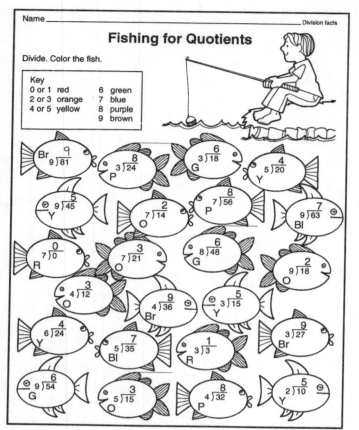

Br $9\overline{)81}$ → 9
P $3\overline{)24}$ → 8
G $3\overline{)18}$ → 6
Y $5\overline{)20}$ → 4
Y $9\overline{)45}$ → 5
O $7\overline{)14}$ → 2
P $7\overline{)56}$ → 8
Bl $9\overline{)63}$ → 7
R $7\overline{)0}$ → 0
O $7\overline{)21}$ → 3
G $8\overline{)48}$ → 6
O $9\overline{)18}$ → 2
$4\overline{)12}$ → 3
Br $4\overline{)36}$ → 9
Y $3\overline{)15}$ → 5
Y $6\overline{)24}$ → 4
Bl $5\overline{)35}$ → 7
R $3\overline{)3}$ → 1
Br $3\overline{)27}$ → 9
G $9\overline{)54}$ → 6
O $5\overline{)15}$ → 3
P $4\overline{)32}$ → 8
Y $2\overline{)10}$ → 5

Page 61

Puzzle Pieces

Division facts

Match.

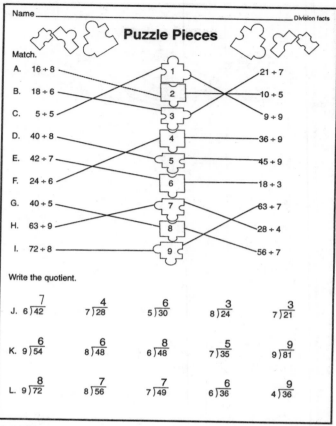

A. 16 ÷ 8
B. 18 ÷ 6
C. 5 ÷ 5
D. 40 ÷ 8
E. 42 ÷ 7
F. 24 ÷ 6
G. 40 ÷ 5
H. 63 ÷ 9
I. 72 ÷ 8

1   21 ÷ 7
2   10 ÷ 5
3   9 ÷ 9
4   36 ÷ 9
5   45 ÷ 9
6   18 ÷ 3
7   63 ÷ 7
8   28 ÷ 4
9   56 ÷ 7

Write the quotient.

J. $6\overline{)42}$ = 7 $7\overline{)28}$ = 4 $5\overline{)30}$ = 6 $8\overline{)24}$ = 3 $7\overline{)21}$ = 3

K. $9\overline{)54}$ = 6 $8\overline{)48}$ = 6 $6\overline{)48}$ = 8 $7\overline{)35}$ = 5 $9\overline{)81}$ = 9

L. $9\overline{)72}$ = 8 $8\overline{)56}$ = 7 $7\overline{)49}$ = 7 $6\overline{)36}$ = 6 $4\overline{)36}$ = 9

Page 62

Family Albums

Fact families

Multiply or divide.
Write the numbers for each fact family in the album label.

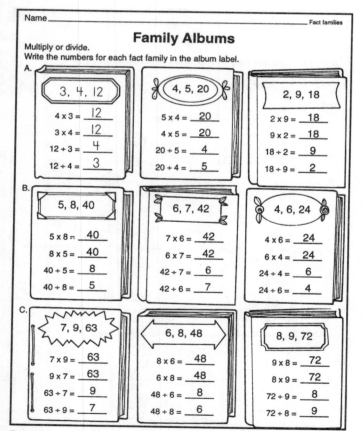

A.
3, 4, 12
4 x 3 = 12
3 x 4 = 12
12 ÷ 3 = 4
12 ÷ 4 = 3

4, 5, 20
5 x 4 = 20
4 x 5 = 20
20 ÷ 5 = 4
20 ÷ 4 = 5

2, 9, 18
2 x 9 = 18
9 x 2 = 18
18 ÷ 2 = 9
18 ÷ 9 = 2

B.
5, 8, 40
5 x 8 = 40
8 x 5 = 40
40 ÷ 5 = 8
40 ÷ 8 = 5

6, 7, 42
7 x 6 = 42
6 x 7 = 42
42 ÷ 7 = 6
42 ÷ 6 = 7

4, 6, 24
4 x 6 = 24
6 x 4 = 24
24 ÷ 4 = 6
24 ÷ 6 = 4

C.
7, 9, 63
7 x 9 = 63
9 x 7 = 63
63 ÷ 7 = 9
63 ÷ 9 = 7

6, 8, 48
8 x 6 = 48
6 x 8 = 48
48 ÷ 6 = 8
48 ÷ 8 = 6

8, 9, 72
9 x 8 = 72
8 x 9 = 72
72 ÷ 9 = 8
72 ÷ 8 = 9

Page 63

Team Jerseys

Fact families

Write a fact family for each group of numbers.

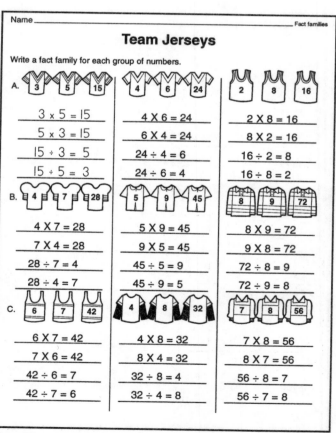

A. **3 5 15**
3 x 5 = 15
5 x 3 = 15
15 ÷ 3 = 5
15 ÷ 5 = 3

4 6 24
4 X 6 = 24
6 X 4 = 24
24 ÷ 4 = 6
24 ÷ 6 = 4

2 8 16
2 X 8 = 16
8 X 2 = 16
16 ÷ 2 = 8
16 ÷ 8 = 2

B. **4 7 28**
4 X 7 = 28
7 X 4 = 28
28 ÷ 7 = 4
28 ÷ 4 = 7

5 9 45
5 X 9 = 45
9 X 5 = 45
45 ÷ 5 = 9
45 ÷ 9 = 5

8 9 72
8 X 9 = 72
9 X 8 = 72
72 ÷ 8 = 9
72 ÷ 9 = 8

C. **6 7 42**
6 X 7 = 42
7 X 6 = 42
42 ÷ 6 = 7
42 ÷ 7 = 6

4 8 32
4 X 8 = 32
8 X 4 = 32
32 ÷ 8 = 4
32 ÷ 4 = 8

7 8 56
7 X 8 = 56
8 X 7 = 56
56 ÷ 8 = 7
56 ÷ 7 = 8

Page 64

Answer Key

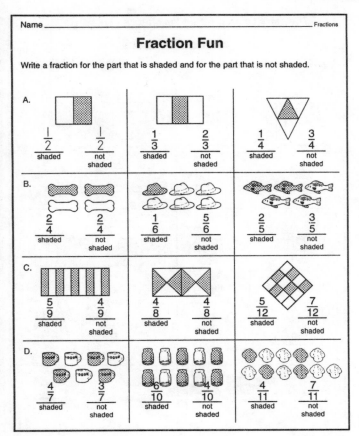

Fraction Fun

Page 65

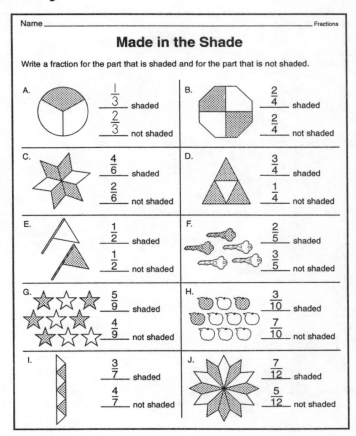

Made in the Shade

Page 66

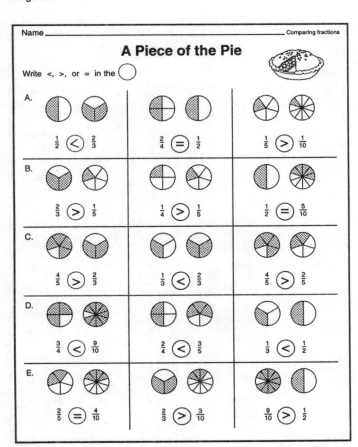

A Piece of the Pie

Page 67

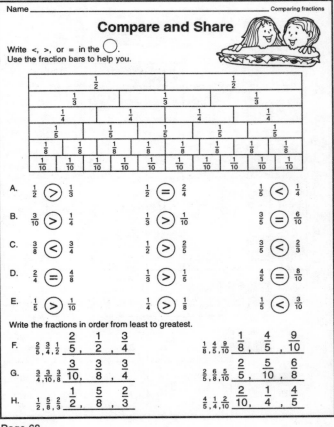

Compare and Share

Page 68

FS-32070 Third Grade Math Review

Answer Key

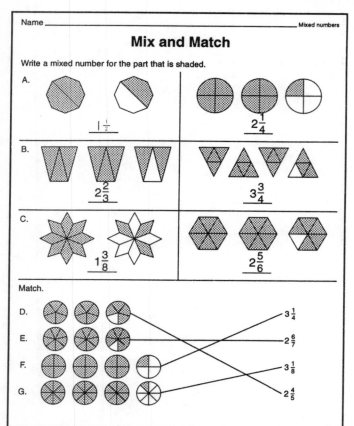

Page 69

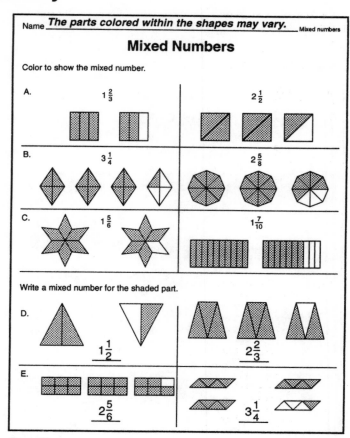

Page 70

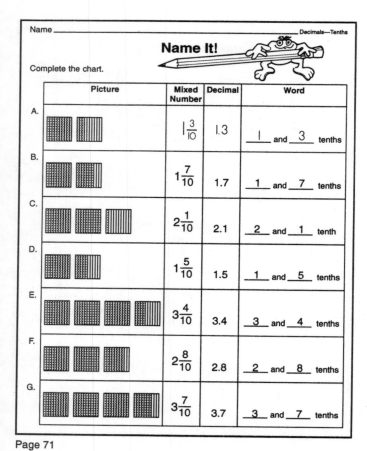

Page 71

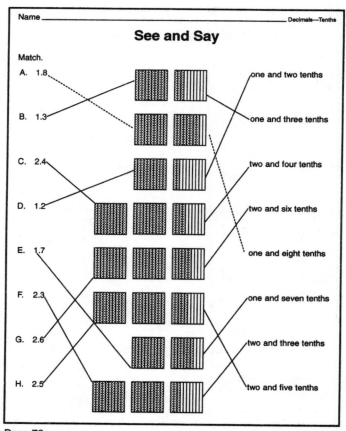

Page 72

FS-32070 Third Grade Math Review

Size It Up

Name _____

Comparing and ordering decimals

Compare. Write > or < in the ◯

A. 1.1 < 1.3   3.5 > 2.7   4.5 < 4.7

B. 4.9 < 5.3   2.7 > 1.9   3.5 > 2.8

C. 2.2 > 1.1   5.6 < 5.7   9.2 > 8.9

D. 8.0 > 7.8   2.6 < 6.2   1.9 < 2.0

E. 2.4 > 2.0   3.9 > 2.5   1.8 > 1.3

F. 3.5 > 3.1   4.2 > 3.8   7.1 < 7.2

G. 6.2 < 7.0   8.3 < 8.9   9.5 > 9.4

Write the decimals from least to greatest.

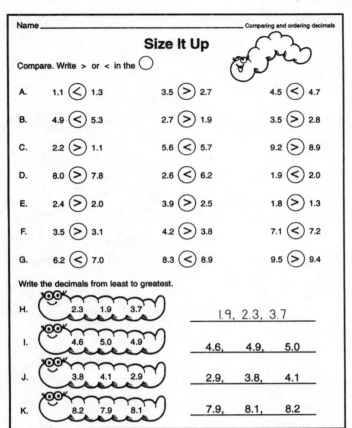

H. 2.3 1.9 3.7   1.9, 2.3, 3.7

I. 4.6 5.0 4.9   4.6,   4.9,   5.0

J. 3.8 4.1 2.9   2.9,   3.8,   4.1

K. 8.2 7.9 8.1   7.9,   8.1,   8.2

Page 73

Leaping Along

Name _____

Comparing and ordering decimals

Compare. Write < or > in the ◯

A. 3.7 > 3.4   2.5 < 2.9   4.2 < 5.2

B. 6.1 < 6.8   4.6 > 3.6   8.9 < 9.5

C. 2.8 < 3.1   1.7 > 1.1   4.3 > 4.0

D. 3.5 < 3.6   9.8 > 8.9   7.3 > 6.8

E. 3.1 < 3.8   1.4 < 2.6   3.5 < 3.9

F. 4.8 < 5.1   6.2 < 6.5   5.7 < 5.8

G. 6.3 > 5.8   7.2 < 7.3   8.5 < 9.2

H. 8.9 < 9.3   7.4 > 6.9   4.6 < 5.3

Write the decimals from least to greatest.

I. 2.3, 2.0, 2.4   2.0   2.3   2.4

J. 6.5, 6.2, 6.7   6.2   6.5   6.7

K. 5.0, 5.1, 4.9   4.9   5.0   5.1

L. 7.8, 7.9, 7.7   7.7   7.8   7.9

Page 74

Decimal Sums and Differences

Name _____

Adding and subtracting decimals

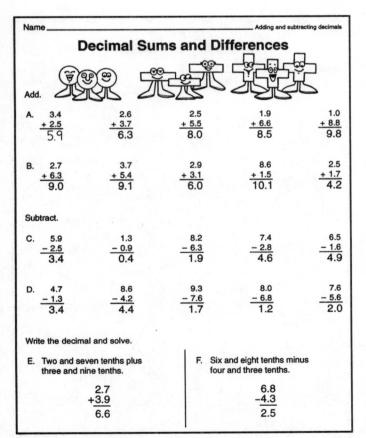

Add.

A. 3.4   2.6   2.5   1.9   1.0
  + 2.5   + 3.7   + 5.5   + 6.6   + 8.8
   5.9   6.3   8.0   8.5   9.8

B. 2.7   3.7   2.9   8.6   2.5
  + 6.3   + 5.4   + 3.1   + 1.5   + 1.7
   9.0   9.1   6.0   10.1   4.2

Subtract.

C. 5.9   1.3   8.2   7.4   6.5
  − 2.5   − 0.9   − 6.3   − 2.8   − 1.6
   3.4   0.4   1.9   4.6   4.9

D. 4.7   8.6   9.3   8.0   7.6
  − 1.3   − 4.2   − 7.6   − 6.8   − 5.6
   3.4   4.4   1.7   1.2   2.0

Write the decimal and solve.

E. Two and seven tenths plus three and nine tenths.

  2.7
 +3.9
  6.6

F. Six and eight tenths minus four and three tenths.

  6.8
 −4.3
  2.5

Page 75

Adding and Subtracting Decimals

Name _____

Adding and subtracting decimals

Write the decimal and solve.

A. Two and three tenths plus five and nine tenths.

  2.3
 +5.9
  8.2

B. Six and eight tenths plus one and seven tenths.

  6.8
 +1.7
  8.5

C. Six and five tenths minus four and two tenths.

  6.5
 −4.2
  2.3

D. Three and two tenths minus two and three tenths.

  3.2
 −2.3
  0.9

Add.

E. 4.4   3.5   6.5   9.3   6.7
  + 1.7   + 2.9   + 1.3   + 0.5   + 2.9
   6.1   6.4   7.8   9.8   9.6

F. 8.4   7.8   4.3   8.5   3.6
  + 1.3   + 1.9   + 2.7   + 1.0   + 4.6
   9.7   9.7   7.0   9.5   8.2

Subtract.

G. 9.8   4.7   8.1   4.3   9.3
  − 5.2   − 2.9   − 5.1   − 3.9   − 6.8
   4.6   1.8   3.0   0.4   2.5

H. 6.0   7.5   5.3   8.6   7.2
  − 1.5   − 5.0   − 2.7   − 8.0   − 4.8
   4.5   2.5   2.6   0.6   2.4

Page 76

FS-32070 Third Grade Math Review

Answer Key

Name

Telling time—Minutes

Time for Fun

Draw a line from each clock to the clock with the matching time.

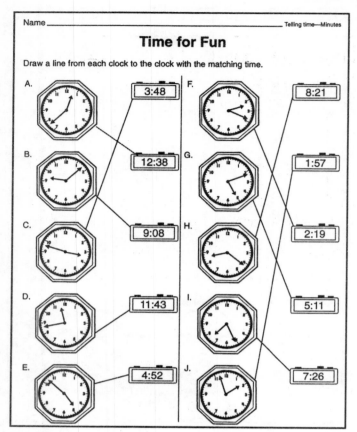

Page 77

Name

Telling time—Minutes

Party Time

Write the time.

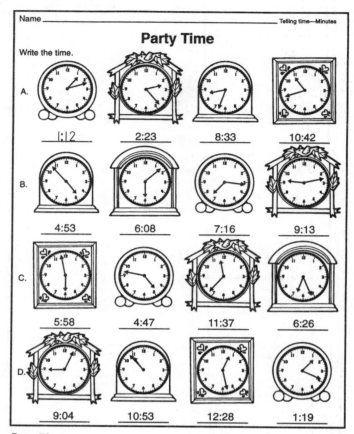

A. 1:12   2:23   8:33   10:42

B. 4:53   6:08   7:16   9:13

C. 5:58   4:47   11:37   6:26

D. 9:04   10:53   12:28   1:19

Page 78

Name

Elapsed time

Timers

Write how much time has passed.

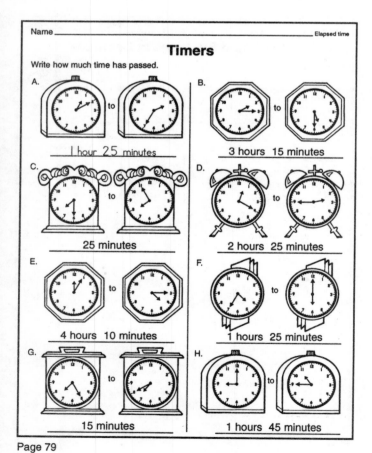

A. 1 hour 25 minutes   B. 3 hours 15 minutes
C. 25 minutes   D. 2 hours 25 minutes
E. 4 hours 10 minutes   F. 1 hours 25 minutes
G. 15 minutes   H. 1 hours 45 minutes

Page 79

Name

Elapsed time

How Long Will It Take?

Write how much time has passed.

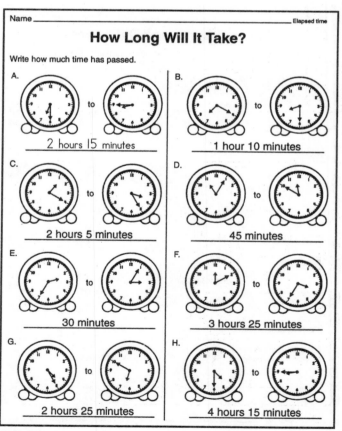

A. 2 hours 15 minutes   B. 1 hour 10 minutes
C. 2 hours 5 minutes   D. 45 minutes
E. 30 minutes   F. 3 hours 25 minutes
G. 2 hours 25 minutes   H. 4 hours 15 minutes

Page 80

FS-32070 Third Grade Math Review

Answer Key

Page 83 — Shopping Spree (Adding and subtracting money)

Add.

(example) $2.83 + 2.87 = $6.12

A.	$2.83 + 1.97 = $4.80	$4.56 + 1.28 = $5.84	$6.80 + 1.49 = $8.29	$3.27 + 2.58 = $5.85
B.	$4.65 + 1.83 = $6.48	$3.79 + 4.44 = $8.23	$2.15 + 3.97 = $6.12	$5.63 + 3.79 = $9.42	$7.18 + 2.76 = $9.94

Subtract.

C.	$2.98 − 1.49 = $1.49	$3.75 − 2.90 = $0.85	$4.02 − 3.98 = $0.04	$7.94 − 5.47 = $2.47
D.	$6.73 − 2.85 = $3.88	$5.84 − 2.95 = $2.89	$9.05 − 3.49 = $5.56	$7.37 − 3.64 = $3.73

Write a number sentence and solve.

E. Serene bought a cap for $5.15 and a book for $2.95. How much did she spend in all?
  $5.15 + 2.95 = $8.10

F. Paul had $5.00. He spent $2.97 on stickers. How much money does he have left?
  $5.00 − 2.97 = $2.03

Page 82 — Grocery Shopping (Making change)

Mark the coins you would get for change. Use the fewest coins you can. Write the amount of change.

	Item Bought	Money Paid	$1	50¢	25¢	10¢	5¢	1¢	Amount of Change
A.	apple $0.34	$0.50				1	1	1	16¢
B.	$2.44	$3.00		1			1	1	56¢
C.	$0.76	$1.00				2		4	24¢
D.	$1.69	$2.00			1		1	1	31¢
E.	$1.55	$2.00			1	2			45¢
F.	$3.98	$5.00	1					2	$1.02

Page 81 — What's the Change? (Making change)

Cross out the change. Write the amount.

	Item Bought	Money Paid	Coins for Change	Amount of Change
A.	pencil $0.39		(coins crossed out)	61¢
B.	notebook $1.27		(coins crossed out)	73¢
C.	$2.98		(coins crossed out)	$2.02
D.	$0.85		(coins crossed out)	15¢
E.	eraser $1.07		(coins crossed out)	93¢
F.	glue $1.53		(coins crossed out)	47¢

Page 84 — Money in the Bank (Adding and subtracting money)

Add or subtract.

A.	$2.29 + 6.87 = $9.16	$4.53 − 3.98 = $0.55	$5.51 + 2.99 = $8.50	$6.84 − 3.49 = $3.35	$3.69 + 4.73 = $8.42
B.	$2.86 + 4.25 = $7.11	$5.92 − 1.85 = $4.07	$4.95 − 4.07 = $0.88	$1.57 + 4.98 = $6.55	$8.48 − 3.72 = $4.76
C.	$9.47 − 3.89 = $5.58	$4.02 − 1.95 = $2.07	$3.42 + 4.57 = $7.99	$8.35 − 3.94 = $4.41	$6.09 + 5.25 = $11.34
D.	$8.00 − 4.49 = $3.51	$2.57 + 3.98 = $6.55	$8.42 − 1.79 = $6.63	$2.68 + 3.99 = $6.67	$7.32 − 3.21 = $4.11
E.	$2.48 + 4.29 = $6.77	$7.03 − 3.97 = $3.06	$5.15 − 2.84 = $2.31	$2.89 + 3.47 = $6.36	$6.45 − 4.32 = $2.13

Write a number sentence and solve.

F. Marc had $5.95 in his bank. Then he added $2.15 to it. How much is in Marc's bank now?
  $5.95 + 2.15 = $8.10

G. Rosie had $9.00. She spent $2.49 on a toy boat. How much does Rosie have now?
  $9.00 − 2.49 = $6.51

Page 85 — Special Tens (Multiplying by multiples of 10)

Multiply.

A.	80 × 6 = 480	30 × 5 = 150	60 × 2 = 120	50 × 4 = 200	40 × 6 = 240
B.	40 × 4 = 160	50 × 7 = 350	20 × 3 = 60	70 × 3 = 210	10 × 8 = 80
C.	80 × 4 = 320	90 × 4 = 360	10 × 5 = 50	20 × 7 = 140	30 × 9 = 270
D.	40 × 7 = 280	50 × 6 = 300	70 × 8 = 560	60 × 6 = 360	70 × 2 = 140
E.	90 × 6 = 540	80 × 8 = 640	30 × 8 = 240	40 × 5 = 200	60 × 8 = 480
F.	70 × 7 = 490	90 × 7 = 630	60 × 9 = 540	30 × 7 = 210	90 × 9 = 810

Page 86 — Wonderful Hundreds (Multiplying by multiples of 100)

Multiply.

A. 5 × 1 = 5 ; 5 × 10 = 50 ; 5 × 100 = 500 ; 7 × 7 = 49 ; 7 × 70 = 490 ; 7 × 700 = 4,900 ; 3 × 5 = 15 ; 3 × 50 = 150 ; 3 × 500 = 1,500

B. 800 × 2 = 1,600 ; 900 × 6 = 5,400 ; 200 × 4 = 800 ; 500 × 2 = 1,000 ; 300 × 3 = 900

C. 600 × 3 = 1,800 ; 700 × 3 = 2,100 ; 200 × 4 = 800 ; 300 × 4 = 1,200 ; 400 × 7 = 2,800

D. 800 × 3 = 2,400 ; 100 × 5 = 500 ; 500 × 4 = 2,000 ; 600 × 4 = 2,400 ; 900 × 3 = 2,700

E. 200 × 7 = 1,400 ; 400 × 4 = 1,600 ; 300 × 6 = 1,800 ; 900 × 9 = 8,100

F. 700 × 4 = 2,800 ; 500 × 7 = 3,500 ; 600 × 8 = 4,800 ; 500 × 5 = 2,500 ; 200 × 2 = 400 ; 100 × 8 = 800

FS-32070 Third Grade Math Review

Strawberry Patch
Multiplying two-digit by one-digit numbers—No regrouping

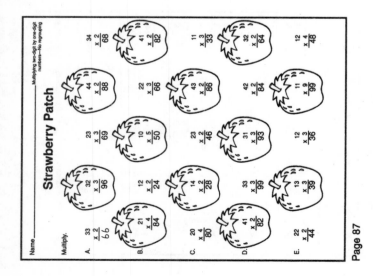

Multiply.

A. 33 ×2 = 66; 44 ×2 = 88; 23 ×3 = 69; 32 ×3 = 96; 34 ×2 = 68
B. 21 ×4 = 84; 22 ×3 = 66; 10 ×5 = 50; 12 ×2 = 24; 41 ×2 = 82
C. 20 ×4 = 80; 43 ×2 = 86; 23 ×2 = 46; 14 ×2 = 28; 11 ×3 = 33
D. 41 ×2 = 82; 42 ×2 = 84; 31 ×3 = 93; 33 ×3 = 99; 32 ×2 = 64
E. 22 ×2 = 44; 13 ×3 = 39; 12 ×3 = 36; 11 ×9 = 99; 12 ×4 = 48

Page 87

Multiplication Paint
Multiplying two-digit by one-digit numbers—No regrouping

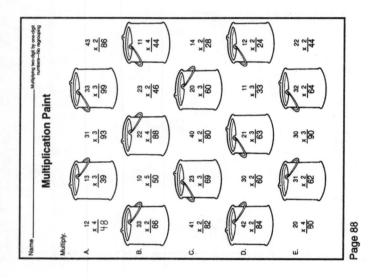

Multiply.

A. 12 ×4 = 48; 13 ×3 = 39; 31 ×3 = 93; 33 ×3 = 99; 43 ×2 = 86
B. 33 ×2 = 66; 10 ×5 = 50; 22 ×4 = 88; 23 ×2 = 46; 11 ×4 = 44
C. 41 ×2 = 82; 23 ×3 = 69; 40 ×2 = 80; 20 ×3 = 60; 14 ×2 = 28
D. 42 ×2 = 84; 30 ×2 = 60; 21 ×3 = 63; 32 ×2 = 64; 12 ×2 = 24
E. 20 ×4 = 80; 31 ×2 = 62; 30 ×3 = 90; 22 ×2 = 44

Page 88

A Buried Treasure
Multiplying two-digit by one-digit numbers greater than 70.

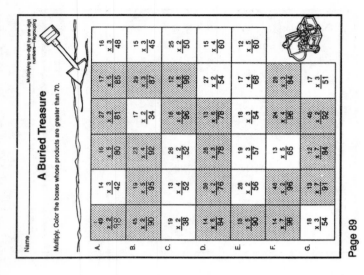

Multiply. Color the boxes whose products are greater than 70.

A. 49 ×2 = 98; 14 ×3 = 42; 16 ×5 = 80; 27 ×3 = 81; 17 ×5 = 85; 16 ×3 = 48
B. 45 ×2 = 90; 19 ×5 = 95; 23 ×4 = 92; 17 ×2 = 34; 29 ×3 = 87; 15 ×3 = 45
C. 19 ×2 = 38; 13 ×4 = 52; 26 ×2 = 52; 16 ×6 = 96; 12 ×8 = 96; 25 ×2 = 50
D. 14 ×6 = 84; 38 ×2 = 76; 26 ×3 = 78; 13 ×6 = 78; 27 ×2 = 54; 15 ×4 = 60
E. 18 ×5 = 90; 28 ×2 = 56; 19 ×3 = 57; 18 ×3 = 54; 17 ×4 = 68; 12 ×5 = 60
F. 14 ×7 = 98; 46 ×2 = 96; 13 ×5 = 65; 24 ×4 = 96; 20 ×3 = 84
G. 18 ×3 = 54; 13 ×7 = 91; 12 ×7 = 84; 46 ×2 = 92; 17 ×3 = 51

Page 89

Hooked on Multiplication
Multiplying two-digit by one-digit numbers—Regrouping

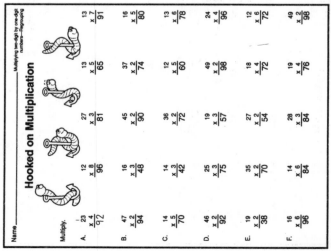

Multiply.

A. 23 ×4 = 92; 16 ×6 = 96; 14 ×3 = 42; 27 ×3 = 81; 13 ×7 = 91
B. 47 ×2 = 94; 35 ×2 = 70; 25 ×3 = 75; 45 ×2 = 90; 13 ×5 = 65
C. 14 ×5 = 70; 14 ×3 = 42; 36 ×2 = 72; 37 ×2 = 74; 12 ×5 = 60
D. 46 ×2 = 92; 19 ×2 = 38; 19 ×3 = 57; 13 ×6 = 78
E. 19 ×2 = 38; 27 ×2 = 54; 18 ×4 = 72; 24 ×4 = 96
F. 16 ×6 = 96; 14 ×6 = 84; 28 ×3 = 84; 12 ×6 = 72; 19 ×4 = 76; 49 ×2 = 98

Page 90

Look Out for Multiplication
Multiplying two-digit by one-digit numbers—Regrouping

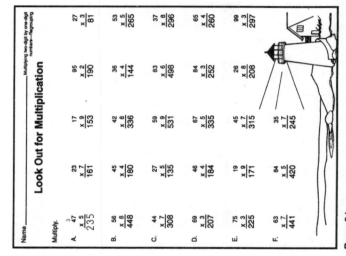

Multiply.

A. 47 ×5 = 235; 23 ×7 = 161; 17 ×9 = 153; 95 ×2 = 190; 27 ×3 = 81
B. 56 ×8 = 448; 45 ×4 = 180; 42 ×8 = 336; 36 ×4 = 144; 53 ×5 = 265
C. 44 ×7 = 308; 27 ×5 = 135; 59 ×9 = 531; 83 ×6 = 498; 37 ×8 = 296
D. 69 ×3 = 207; 46 ×4 = 184; 67 ×5 = 335; 84 ×3 = 252; 65 ×4 = 260
E. 75 ×3 = 225; 19 ×9 = 171; 45 ×7 = 315; 26 ×8 = 208; 99 ×3 = 297
F. 63 ×7 = 441; 84 ×5 = 420; 35 ×7 = 245

Page 91

Croaking About Multiplication
Multiplying two-digit by one-digit numbers—Regrouping

Multiply.

A. 37 ×6 = 222; 27 ×4 = 108; 46 ×5 = 230; 35 ×7 = 245; 56 ×4 = 224
B. 84 ×8 = 672; 19 ×9 = 171; 28 ×6 = 168; 39 ×7 = 273; 67 ×4 = 268
C. 48 ×5 = 240; 17 ×9 = 153; 38 ×7 = 266; 65 ×4 = 260; 59 ×5 = 295
D. 68 ×2 = 136; 79 ×3 = 237; 27 ×9 = 243; 45 ×7 = 315; 34 ×8 = 272
E. 95 ×5 = 475; 87 ×3 = 261; 47 ×5 = 235; 16 ×8 = 128; 77 ×6 = 462
F. 36 ×8 = 288; 89 ×3 = 267; 58 ×6 = 348; 34 ×9 = 306; 78 ×4 = 312

Page 92

FS-32070 Third Grade Math Review

Multiplication Wizard (Page 94)

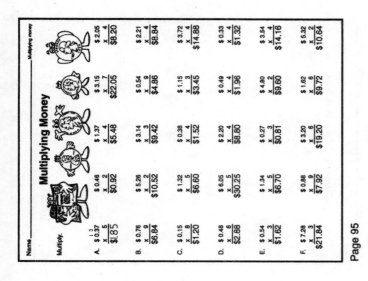

Name _____ Multiplying three-digit by one-digit numbers—Regrouping

Multiply.

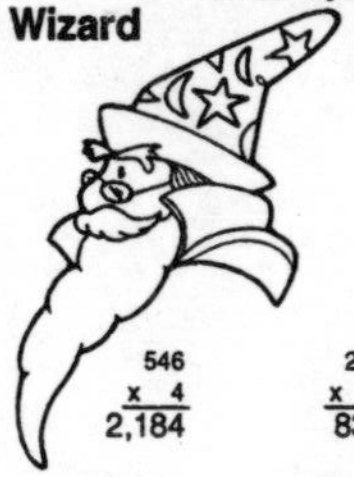

A. $\begin{array}{r} {\scriptstyle 1} \\ 351 \\ \underline{\times\ 2} \\ 702 \end{array}$	$\begin{array}{r} 372 \\ \underline{\times\ 3} \\ 1{,}116 \end{array}$	$\begin{array}{r} 124 \\ \underline{\times\ 3} \\ 372 \end{array}$	$\begin{array}{r} 546 \\ \underline{\times\ 4} \\ 2{,}184 \end{array}$	$\begin{array}{r} 208 \\ \underline{\times\ 4} \\ 832 \end{array}$
B. $\begin{array}{r} 103 \\ \underline{\times\ 9} \\ 927 \end{array}$	$\begin{array}{r} 249 \\ \underline{\times\ 4} \\ 996 \end{array}$	$\begin{array}{r} 377 \\ \underline{\times\ 2} \\ 754 \end{array}$		$\begin{array}{r} 158 \\ \underline{\times\ 5} \\ 790 \end{array}$
C. $\begin{array}{r} 382 \\ \underline{\times\ 3} \\ 1{,}146 \end{array}$	$\begin{array}{r} 417 \\ \underline{\times\ 2} \\ 834 \end{array}$	$\begin{array}{r} 126 \\ \underline{\times\ 4} \\ 504 \end{array}$	$\begin{array}{r} 238 \\ \underline{\times\ 3} \\ 714 \end{array}$	
D. $\begin{array}{r} 206 \\ \underline{\times\ 3} \\ 618 \end{array}$	$\begin{array}{r} 324 \\ \underline{\times\ 3} \\ 972 \end{array}$	$\begin{array}{r} 168 \\ \underline{\times\ 4} \\ 672 \end{array}$	$\begin{array}{r} 295 \\ \underline{\times\ 3} \\ 885 \end{array}$	$\begin{array}{r} 128 \\ \underline{\times\ 3} \\ 384 \end{array}$
E. $\begin{array}{r} 135 \\ \underline{\times\ 4} \\ 540 \end{array}$	$\begin{array}{r} 175 \\ \underline{\times\ 5} \\ 875 \end{array}$	$\begin{array}{r} 309 \\ \underline{\times\ 2} \\ 618 \end{array}$	$\begin{array}{r} 218 \\ \underline{\times\ 4} \\ 872 \end{array}$	$\begin{array}{r} 408 \\ \underline{\times\ 2} \\ 816 \end{array}$
F. $\begin{array}{r} 196 \\ \underline{\times\ 4} \\ 784 \end{array}$	$\begin{array}{r} 319 \\ \underline{\times\ 2} \\ 638 \end{array}$	$\begin{array}{r} 256 \\ \underline{\times\ 3} \\ 768 \end{array}$	$\begin{array}{r} 165 \\ \underline{\times\ 4} \\ 660 \end{array}$	$\begin{array}{r} 127 \\ \underline{\times\ 6} \\ 762 \end{array}$

Page 94

Multiplying Money (Page 95)

Name _____ Multiplying money

Multiply.

A. $\begin{array}{r} {\scriptstyle 1\ 3} \\ \$0.37 \\ \underline{\times\ 5} \\ \$1.85 \end{array}$	$\begin{array}{r} \$0.46 \\ \underline{\times\ 2} \\ \$0.92 \end{array}$	$\begin{array}{r} \$1.37 \\ \underline{\times\ 4} \\ \$5.48 \end{array}$	$\begin{array}{r} \$3.15 \\ \underline{\times\ 7} \\ \$22.05 \end{array}$	$\begin{array}{r} \$2.05 \\ \underline{\times\ 4} \\ \$8.20 \end{array}$
B. $\begin{array}{r} \$0.76 \\ \underline{\times\ 9} \\ \$6.84 \end{array}$	$\begin{array}{r} \$5.26 \\ \underline{\times\ 2} \\ \$10.52 \end{array}$	$\begin{array}{r} \$3.14 \\ \underline{\times\ 3} \\ \$9.42 \end{array}$	$\begin{array}{r} \$0.54 \\ \underline{\times\ 9} \\ \$4.86 \end{array}$	$\begin{array}{r} \$2.21 \\ \underline{\times\ 4} \\ \$8.84 \end{array}$
C. $\begin{array}{r} \$0.15 \\ \underline{\times\ 8} \\ \$1.20 \end{array}$	$\begin{array}{r} \$1.32 \\ \underline{\times\ 5} \\ \$6.60 \end{array}$	$\begin{array}{r} \$0.38 \\ \underline{\times\ 4} \\ \$1.52 \end{array}$	$\begin{array}{r} \$1.15 \\ \underline{\times\ 3} \\ \$3.45 \end{array}$	$\begin{array}{r} \$3.72 \\ \underline{\times\ 4} \\ \$14.88 \end{array}$
D. $\begin{array}{r} \$0.48 \\ \underline{\times\ 6} \\ \$2.88 \end{array}$	$\begin{array}{r} \$6.05 \\ \underline{\times\ 5} \\ \$30.25 \end{array}$	$\begin{array}{r} \$2.20 \\ \underline{\times\ 4} \\ \$8.80 \end{array}$	$\begin{array}{r} \$0.49 \\ \underline{\times\ 4} \\ \$1.96 \end{array}$	$\begin{array}{r} \$0.33 \\ \underline{\times\ 4} \\ \$1.32 \end{array}$
E. $\begin{array}{r} \$0.54 \\ \underline{\times\ 3} \\ \$1.62 \end{array}$	$\begin{array}{r} \$1.34 \\ \underline{\times\ 5} \\ \$6.70 \end{array}$	$\begin{array}{r} \$0.27 \\ \underline{\times\ 3} \\ \$0.81 \end{array}$	$\begin{array}{r} \$4.80 \\ \underline{\times\ 2} \\ \$9.60 \end{array}$	$\begin{array}{r} \$3.54 \\ \underline{\times\ 4} \\ \$14.16 \end{array}$
F. $\begin{array}{r} \$7.28 \\ \underline{\times\ 3} \\ \$21.84 \end{array}$	$\begin{array}{r} \$0.88 \\ \underline{\times\ 9} \\ \$7.92 \end{array}$	$\begin{array}{r} \$3.20 \\ \underline{\times\ 6} \\ \$19.20 \end{array}$	$\begin{array}{r} \$1.62 \\ \underline{\times\ 6} \\ \$9.72 \end{array}$	$\begin{array}{r} \$5.32 \\ \underline{\times\ 2} \\ \$10.64 \end{array}$

Page 95

Skating Through Multiplication (Page 93)

Name _____ Multiplying three-digit by one-digit numbers—Regrouping

Multiply.

A. $\begin{array}{r} {\scriptstyle 3} \\ 116 \\ \underline{\times\ 6} \\ 696 \end{array}$	$\begin{array}{r} 161 \\ \underline{\times\ 5} \\ 805 \end{array}$	$\begin{array}{r} 125 \\ \underline{\times\ 3} \\ 375 \end{array}$	$\begin{array}{r} 238 \\ \underline{\times\ 2} \\ 476 \end{array}$	$\begin{array}{r} 243 \\ \underline{\times\ 3} \\ 729 \end{array}$
B. $\begin{array}{r} 484 \\ \underline{\times\ 2} \\ 968 \end{array}$	$\begin{array}{r} 107 \\ \underline{\times\ 9} \\ 963 \end{array}$	$\begin{array}{r} 171 \\ \underline{\times\ 5} \\ 855 \end{array}$	$\begin{array}{r} 216 \\ \underline{\times\ 4} \\ 864 \end{array}$	$\begin{array}{r} 217 \\ \underline{\times\ 4} \\ 868 \end{array}$
C. $\begin{array}{r} 106 \\ \underline{\times\ 5} \\ 530 \end{array}$	$\begin{array}{r} 246 \\ \underline{\times\ 2} \\ 492 \end{array}$	$\begin{array}{r} 219 \\ \underline{\times\ 4} \\ 876 \end{array}$	$\begin{array}{r} 119 \\ \underline{\times\ 5} \\ 595 \end{array}$	$\begin{array}{r} 283 \\ \underline{\times\ 3} \\ 849 \end{array}$
D. $\begin{array}{r} 108 \\ \underline{\times\ 5} \\ 540 \end{array}$	$\begin{array}{r} 231 \\ \underline{\times\ 4} \\ 924 \end{array}$	$\begin{array}{r} 429 \\ \underline{\times\ 2} \\ 858 \end{array}$	$\begin{array}{r} 407 \\ \underline{\times\ 2} \\ 814 \end{array}$	$\begin{array}{r} 272 \\ \underline{\times\ 3} \\ 816 \end{array}$
E. $\begin{array}{r} 427 \\ \underline{\times\ 2} \\ 854 \end{array}$	$\begin{array}{r} 109 \\ \underline{\times\ 6} \\ 654 \end{array}$	$\begin{array}{r} 218 \\ \underline{\times\ 4} \\ 872 \end{array}$	$\begin{array}{r} 329 \\ \underline{\times\ 3} \\ 987 \end{array}$	$\begin{array}{r} 124 \\ \underline{\times\ 3} \\ 372 \end{array}$
F. $\begin{array}{r} 724 \\ \underline{\times\ 3} \\ 2{,}172 \end{array}$	$\begin{array}{r} 205 \\ \underline{\times\ 4} \\ 820 \end{array}$	$\begin{array}{r} 352 \\ \underline{\times\ 4} \\ 1{,}408 \end{array}$	$\begin{array}{r} 627 \\ \underline{\times\ 3} \\ 1{,}881 \end{array}$	$\begin{array}{r} 283 \\ \underline{\times\ 4} \\ 1{,}132 \end{array}$

Page 93

A Pot of Gold (Page 96)

Name _____ Multiplying money

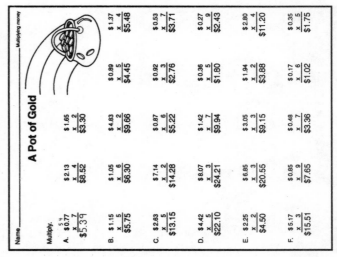

Multiply.

A. $\begin{array}{r} {\scriptstyle 5\ 4} \\ \$0.77 \\ \underline{\times\ 7} \\ \$5.39 \end{array}$	$\begin{array}{r} \$2.13 \\ \underline{\times\ 4} \\ \$8.52 \end{array}$	$\begin{array}{r} \$1.65 \\ \underline{\times\ 2} \\ \$3.30 \end{array}$	$\begin{array}{r} \$0.89 \\ \underline{\times\ 5} \\ \$4.45 \end{array}$	$\begin{array}{r} \$1.37 \\ \underline{\times\ 4} \\ \$5.48 \end{array}$
B. $\begin{array}{r} \$1.15 \\ \underline{\times\ 5} \\ \$5.75 \end{array}$	$\begin{array}{r} \$1.05 \\ \underline{\times\ 6} \\ \$6.30 \end{array}$	$\begin{array}{r} \$4.83 \\ \underline{\times\ 2} \\ \$9.66 \end{array}$	$\begin{array}{r} \$0.92 \\ \underline{\times\ 3} \\ \$2.76 \end{array}$	$\begin{array}{r} \$0.53 \\ \underline{\times\ 7} \\ \$3.71 \end{array}$
C. $\begin{array}{r} \$2.63 \\ \underline{\times\ 5} \\ \$13.15 \end{array}$	$\begin{array}{r} \$7.14 \\ \underline{\times\ 2} \\ \$14.28 \end{array}$	$\begin{array}{r} \$0.87 \\ \underline{\times\ 6} \\ \$5.22 \end{array}$	$\begin{array}{r} \$1.42 \\ \underline{\times\ 7} \\ \$9.94 \end{array}$	$\begin{array}{r} \$0.27 \\ \underline{\times\ 9} \\ \$2.43 \end{array}$
D. $\begin{array}{r} \$4.42 \\ \underline{\times\ 5} \\ \$22.10 \end{array}$	$\begin{array}{r} \$8.07 \\ \underline{\times\ 3} \\ \$24.21 \end{array}$	$\begin{array}{r} \$3.05 \\ \underline{\times\ 3} \\ \$9.15 \end{array}$	$\begin{array}{r} \$0.36 \\ \underline{\times\ 5} \\ \$1.80 \end{array}$	$\begin{array}{r} \$2.80 \\ \underline{\times\ 4} \\ \$11.20 \end{array}$
E. $\begin{array}{r} \$2.25 \\ \underline{\times\ 2} \\ \$4.50 \end{array}$	$\begin{array}{r} \$6.85 \\ \underline{\times\ 3} \\ \$20.55 \end{array}$	$\begin{array}{r} \$0.48 \\ \underline{\times\ 7} \\ \$3.36 \end{array}$	$\begin{array}{r} \$1.94 \\ \underline{\times\ 2} \\ \$3.88 \end{array}$	$\begin{array}{r} \$0.17 \\ \underline{\times\ 6} \\ \$1.02 \end{array}$
F. $\begin{array}{r} \$5.17 \\ \underline{\times\ 3} \\ \$15.51 \end{array}$	$\begin{array}{r} \$0.85 \\ \underline{\times\ 9} \\ \$7.65 \end{array}$		$\begin{array}{r} \$0.35 \\ \underline{\times\ 5} \\ \$1.75 \end{array}$	

Page 96

A Carousel Ride (Page 97)

Name _____ Dividing two-digit numbers by one-digit numbers

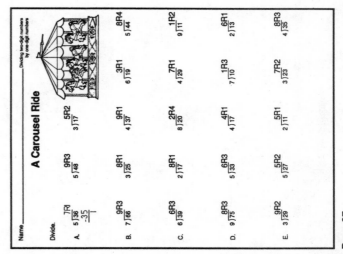

Divide.

A. $\begin{array}{r} 7R1 \\ 5\overline{)36} \\ \underline{-35} \\ 1 \end{array}$	$\begin{array}{r} 9R3 \\ 5\overline{)48} \end{array}$	$\begin{array}{r} 5R2 \\ 3\overline{)17} \end{array}$	$\begin{array}{r} 3R1 \\ 6\overline{)19} \end{array}$	$\begin{array}{r} 8R4 \\ 5\overline{)44} \end{array}$
B. $\begin{array}{r} 9R3 \\ 7\overline{)66} \end{array}$	$\begin{array}{r} 8R1 \\ 3\overline{)25} \end{array}$	$\begin{array}{r} 9R1 \\ 4\overline{)37} \end{array}$	$\begin{array}{r} 7R1 \\ 4\overline{)29} \end{array}$	$\begin{array}{r} 1R2 \\ 9\overline{)11} \end{array}$
C. $\begin{array}{r} 6R3 \\ 6\overline{)39} \end{array}$	$\begin{array}{r} 8R1 \\ 2\overline{)17} \end{array}$	$\begin{array}{r} 2R4 \\ 8\overline{)20} \end{array}$	$\begin{array}{r} 7R1 \\ 4\overline{)29} \end{array}$	$\begin{array}{r} 6R1 \\ 2\overline{)13} \end{array}$
D. $\begin{array}{r} 8R3 \\ 9\overline{)75} \end{array}$	$\begin{array}{r} 6R3 \\ 5\overline{)33} \end{array}$	$\begin{array}{r} 4R1 \\ 4\overline{)17} \end{array}$	$\begin{array}{r} 1R3 \\ 7\overline{)10} \end{array}$	
E. $\begin{array}{r} 9R2 \\ 3\overline{)29} \end{array}$	$\begin{array}{r} 5R2 \\ 5\overline{)27} \end{array}$	$\begin{array}{r} 5R1 \\ 2\overline{)11} \end{array}$	$\begin{array}{r} 7R2 \\ 3\overline{)23} \end{array}$	$\begin{array}{r} 8R3 \\ 4\overline{)35} \end{array}$

Page 97

Let's Divide (Page 98)

Name _____ Dividing two-digit numbers by one-digit numbers

Divide.

A. $\begin{array}{r} 6R4 \\ 5\overline{)34} \\ \underline{-30} \\ 4 \end{array}$	$\begin{array}{r} 7R1 \\ 2\overline{)15} \end{array}$	$\begin{array}{r} 9R1 \\ 3\overline{)28} \end{array}$	$\begin{array}{r} 5R4 \\ 8\overline{)44} \end{array}$	
B. $\begin{array}{r} 2R2 \\ 6\overline{)14} \end{array}$	$\begin{array}{r} 4R2 \\ 7\overline{)30} \end{array}$	$\begin{array}{r} 1R6 \\ 9\overline{)15} \end{array}$	$\begin{array}{r} 7R1 \\ 4\overline{)29} \end{array}$	$\begin{array}{r} 3R2 \\ 7\overline{)23} \end{array}$
C. $\begin{array}{r} 7R2 \\ 5\overline{)37} \end{array}$	$\begin{array}{r} 8R1 \\ 3\overline{)19} \end{array}$	$\begin{array}{r} 6R1 \\ 3\overline{)19} \end{array}$	$\begin{array}{r} 4R1 \\ 6\overline{)25} \end{array}$	$\begin{array}{r} 2R2 \\ 9\overline{)20} \end{array}$
D. $\begin{array}{r} 5R2 \\ 4\overline{)22} \end{array}$	$\begin{array}{r} 4R1 \\ 3\overline{)13} \end{array}$	$\begin{array}{r} 1R1 \\ 9\overline{)10} \end{array}$	$\begin{array}{r} 3R2 \\ 5\overline{)17} \end{array}$	$\begin{array}{r} 6R4 \\ 7\overline{)46} \end{array}$
E. $\begin{array}{r} 5R1 \\ 2\overline{)11} \end{array}$	$\begin{array}{r} 6R2 \\ 3\overline{)20} \end{array}$		$\begin{array}{r} 7R2 \\ 4\overline{)30} \end{array}$	

Page 98

FS-32070 Third Grade Math Review

Answer Key

Step by Step
Dividing two-digit numbers by one-digit numbers

Divide.

A. $14R1$ — $3\overline{)43}$ $12R2$ — $4\overline{)50}$ $12R3$ — $5\overline{)63}$ $19R1$ — $2\overline{)39}$

  $17R1$ — $2\overline{)35}$ $14R2$ — $5\overline{)72}$ $16R3$ — $4\overline{)67}$ $42R1$ — $2\overline{)85}$

B. $12R2$ — $4\overline{)50}$ $27R1$ — $3\overline{)82}$ $16R4$ — $5\overline{)84}$ $12R1$ — $7\overline{)85}$

C. $13R3$ — $5\overline{)68}$ $43R1$ — $2\overline{)87}$ $22R3$ — $4\overline{)91}$ $23R2$ — $4\overline{)94}$

D. $11R4$ — $5\overline{)59}$ $48R1$ — $2\overline{)97}$ $21R1$ — $3\overline{)64}$

Page 100 — A River Ride
Dividing two-digit numbers by one-digit numbers

Divide.

A. $16R3$ — $4\overline{)67}$ $25R1$ — $2\overline{)51}$ $13R1$ — $3\overline{)40}$ $13R4$ — $5\overline{)69}$ $12R5$ — $7\overline{)89}$

B. $12R3$ — $5\overline{)63}$ $11R2$ — $4\overline{)46}$ $14R1$ — $3\overline{)43}$ $32R1$ — $2\overline{)65}$ $14R3$ — $6\overline{)87}$

C. $13R6$ — $7\overline{)97}$ $23R1$ — $4\overline{)93}$ $11R3$ — $5\overline{)58}$ $23R2$ — $3\overline{)71}$ $44R1$ — $2\overline{)89}$

D. $36R1$ — $2\overline{)73}$ $27R1$ — $3\overline{)82}$ $15R3$ — $4\overline{)63}$

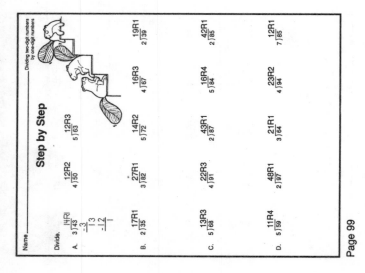

Page 101 — Up, Up, and Away
Dividing three-digit numbers by one-digit numbers

Divide. If the quotient has no remainder, color the section on the hot-air balloon with the matching letter.

A. 42 — $3\overline{)126}$ B. $91R1$ — $5\overline{)456}$ C. $45R1$ — $3\overline{)136}$ D. 88 — $4\overline{)352}$

E. 79 — $2\overline{)158}$ F. $43R2$ — $5\overline{)217}$ G. $85R2$ — $3\overline{)257}$ H. 66 — $4\overline{)264}$

I. $56R1$ — $2\overline{)113}$ J. $28R2$ — $4\overline{)114}$

K. 71 — $5\overline{)355}$ L. 26 — $6\overline{)156}$

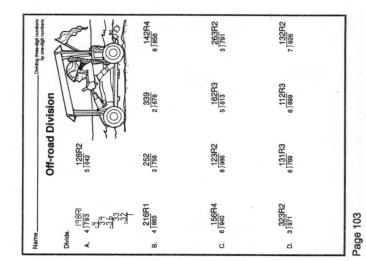

Page 102 — Diamonds Are Forever
Dividing three-digit numbers by one-digit numbers

Divide.

A. 28 — $4\overline{)112}$ B. 21 — $5\overline{)105}$ D. 79 — $3\overline{)237}$ E. $31R3$ — $4\overline{)127}$

F. 30 — $5\overline{)150}$ I. $48R3$ — $4\overline{)195}$ J. 43 — $3\overline{)129}$ N. $21R3$ — $5\overline{)108}$

O. $41R1$ — $5\overline{)206}$ S. $44R3$ — $4\overline{)179}$ T. 41 — $3\overline{)123}$ U. $41R3$ — $4\overline{)167}$

Fill in the correct letter over each answer. Where can you find the largest diamond in the world?

O N   A   B A S E B A L L
F I E L D

O	N	A	B	A	S	E	B	A	L	L
41R1	21R3	28	21	28	44R3	31R3	21	28	43	43

F	I	E	L	D
30	48R3	31R3	43	79

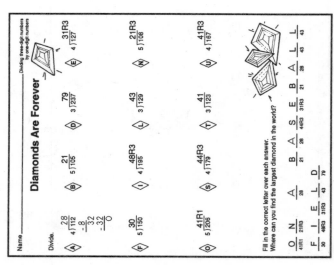

Page 103 — Off-road Division
Dividing three-digit numbers by one-digit numbers

Divide.

A. $198R1$ — $4\overline{)793}$ $128R2$ — $5\overline{)642}$ 339 — $2\overline{)678}$ $142R4$ — $6\overline{)856}$

B. $216R1$ — $4\overline{)865}$ 252 — $3\overline{)756}$ $162R3$ — $5\overline{)813}$ $263R2$ — $3\overline{)791}$

C. $156R4$ — $6\overline{)940}$ $123R2$ — $8\overline{)986}$ $112R3$ — $8\overline{)899}$ $132R2$ — $7\overline{)926}$

D. $323R2$ — $3\overline{)971}$ $131R3$ — $6\overline{)789}$

Page 104 — Brushing Up on Division
Dividing three-digit numbers by one-digit numbers

Divide.

A. $123R3$ — $7\overline{)864}$ $141R3$ — $4\overline{)567}$ 253 — $3\overline{)759}$ $111R1$ — $8\overline{)889}$ $278R1$ — $2\overline{)557}$

B. $137R2$ — $5\overline{)687}$ $132R2$ — $3\overline{)398}$ 297 — $3\overline{)891}$ $463R1$ — $2\overline{)927}$ $156R4$ — $6\overline{)940}$

C. 123 — $8\overline{)984}$ 128 — $6\overline{)768}$

D. $139R3$ — $6\overline{)837}$ $161R3$ — $4\overline{)647}$

FS-32070 Third Grade Math Review